Becoming a Neuropsychologist

John A. Bellone • Ryan Van Patten

Becoming a Neuropsychologist

Advice and Guidance for Students and Trainees

 Springer

John A. Bellone, PhD, ABPP-CN
Perspectives Psychological Services
Fullerton, CA, United States

Ryan Van Patten, PhD
Spaulding Rehabilitation Hospital
Massachusetts General Hospital
Harvard Medical School
Boston, MA, United States

ISBN 978-3-030-63173-4   ISBN 978-3-030-63174-1   (eBook)
https://doi.org/10.1007/978-3-030-63174-1

This Springer imprint is published by the registered company Springer Nature Switzerland AG
The registered company address is: Gewerbestrasse 11, 6330 Cham, Switzerland

Praise for the Book

As our field looks to expand diversity and representation in our ranks, my hat is off to John A. Bellone and Ryan Van Patten for not only extending an invitation to the party, but also for providing a map for how to get here. The opening chapters nicely describe our field and will pique the curiosity of those who are in the process of career exploration. Subsequent chapters are filled with important "how to" tips, including what to study as an undergraduate, how to apply for graduate programs, and then how to cap off training through postdoctoral fellowship and eventually board certification. As such, this is a "now and later" guide, as useful to those who are "neuropsychology-curious" as to those who have already decided to pursue neuropsychology training. Once on the path to becoming a neuropsychologist, the early chapters will be handy to give to family members to read so that they will not think you want to become a "nurse psychologist" (looking at you, Aunt Erica) and can instead share in your enthusiasm for your chosen field. As you move through each phase of training, you can refer back to the book to see what is ahead of you on the road and start to prepare for the next step. The writing is clear and personal, with quotes gathered from a variety of individuals at different career phases. I for one can't wait until the moment I'm interviewing a candidate for fellowship and I ask my standard question of "How did you become interested in neuropsychology?" and get the answer, "Well, there was this book..." Well done gentlemen!
– Kathleen Fuchs, PhD, ABPP, Board-Certified Clinical Neuropsychologist and Associate Professor at the University of Virginia

In *Becoming a Neuropsychologist*, Drs. Bellone and Van Patten have distilled the complexities of neuropsychology career development into an eminently readable and understandable roadmap. Their use of quotes from neuropsychologists and trainees, as well as descriptions of their own experiences makes the book feel personal, even as they enumerate the many steps required and myriad organizations encountered on this journey. This is a great read for anyone with a general interest in the field and a great reference tool for those pursuing this great profession.
– Glenn Smith, PhD, ABPP, Board-Certified Clinical Neuropsychologist and Professor and Chair of Clinical and Health Psychology at the University of Florida

This book is an easy, accessible introduction to the field of neuropsychology. It is written in a very straightforward and practical style that demystifies the career path of clinical neuropsychologists. Drs. Bellone and Van Patten provide a fun and readable guide for students contemplating a career in neuropsychology or for students already embarking on their journey to completing their doctorate. The authors' open and personal accounts of their own journeys toward becoming clinical neuropsychologists read like a fireside chat with a beloved mentor. This book will be a valuable resource for students interested in pursuing a degree in clinical neuropsychology.
– Jenn Davis, PhD, ABPP, Board-Certified Clinical Neuropsychologist and Associate Professor at Brown University and Rhode Island Hospital

Informative, witty, and inspiring. After reading this, I feel re-inspired and excited to be pursuing a career in this field. This is a very well written introduction to the world of neuropsychology.
– Ilex Beltran-Najera, MA, doctoral student of clinical psychology at the University of Houston

Dedications

John: To my grandfathers, John P. Bellone and Salvatore Pennacchio. You taught me that hard work, gratitude, a strong family connection, and overwhelming generosity are the key ingredients to a productive and fulfilling life.

Ryan: To my grandmother, Agnes Van Patten. Your love, self-sacrifice, and dedication to your family will forever inspire me.

Preface

If I could have received one bit of advice during my education, I would have wanted some-one to tell me about neuropsychology sooner.

– Keith Owen Yeates, PhD, ABPP-CN

Choosing a career is simultaneously one of the most important and one of the most challenging decisions you will ever make. It is a pillar of modern adult life, perhaps even on par with romantic partner selection – and we know how tumultuous that process can be! There are numerous relevant factors when considering one's occu-pation. Most people want to enjoy their day-to-day work life while also earning a decent salary, they want to be respected and fulfilled but not burnt out, and they want to have job flexibility without worrying about job security. All of this may seem like a lot to ask for, but we believe that it is possible to find a job that checks all of these boxes. Of course, every career has pros and cons, and no matter what decision we make, we never know the counter-factual – that is, we never know what the outcome would have been if we had selected a different path. So, we suggest that you do the best that you can with the information available to you and read books like the one you have in your hand right now, all in the hopes that you will eventu-ally find that elusive profession that fits your interests and talents like a glove.

The purpose of this book is to introduce you to one career option – neuropsychol-ogy – that may just be that perfect fit for you. Out of so many conceivable paths, why would you choose neuropsychology?… and how would you go about getting there? The answers to these questions lie ahead.

****

Prior to this book, there was no single resource that comprehensively guided interested students through the process of becoming a neuropsychologist.[1] Most people
discovered the profession serendipitously, through a college professor, a mentor, or
a friend. But this isn't good enough. We believe that our field has a great deal to
offer, and we want to do more than simply rely on chance in recruiting the next
generation of neuropsychologists. By providing you with a blueprint for pursuing
this career, we hope to make our field visible and available to anyone who is interested. This way, you don't need luck and/or privilege in order to learn about neuropsychology. We want to attract hardworking, dedicated students from all walks of
life, regardless of their geographic location, socioeconomic status, or cultural background. In fact, it is *precisely* this diversity that will keep the field relevant over the
next few decades as the demographics of the US and other developed nations
undergo major cultural and linguistic shifts.[2,3]

Because the path to becoming a neuropsychologist is long, expensive, and challenging, it is vital that you are fully informed before embarking on this journey.
Indeed, the amount of time, money, and work required to go from high school student or undergraduate to neuropsychologist can feel daunting, and even insurmount-
able. We are here to tell you that this mountain, although tall and steep in certain
locations, can be scaled, and that the process of climbing can be just as rewarding
as the feeling of accomplishment when you reach the peak. With that in mind, we
hope that this book can serve as a North Star to help you navigate to your destination
in the field of neuropsychology.

Part I answers the questions, *What is Neuropsychology?* (Ch. 1), *Why
Neuropsychology?* (Ch. 2), and *Where Do Neuropsychologists Work?* (Ch. 3), and
ends with a discussion of the *Challenges to Working in Neuropsychology* (Ch. 4).
After we have (hopefully) convinced you of the merit of our field, we will move on
to Part II and provide a roadmap on how to move from college to the end point of
working as a full-fledged neuropsychologist. To do this, we have separated chapters
into *Undergraduate Training* (Ch. 5), *Doctoral Training* (Ch. 6), and *Advanced
Training and Practice: Postdoctoral Fellowship and Beyond* (Ch. 7).

Importantly, Parts I and II can stand alone. For those of you who are interested in
learning about neuropsychology but who will not be pursuing a career in this area,
Part I provides a thorough introduction to our profession. On the other hand, if you

---

[1] In fact, the idea for this book arose from questions posed by student listeners of our podcast,
Navigating Neuropsychology (www.NavNeuro.com). Multiple people asked us how to pursue a
career in neuropsychology and we grew tired of responding with, "Unfortunately, there isn't any
one comprehensive resource that describes the process of becoming a neuropsychologist from the
ground up." So, we decided to solve the problem ourselves and write this book.

[2] You can read about the American Academy of Clinical Neuropsychology (AACN) Relevance
2050 Initiative here: www.theaacn.org/relevance-2050/. Also, see www.NavNeuro.com/58.

[3] There are multiple psychology-related resources specifically designed to assist people of diverse
backgrounds. A few include: https://www.apa.org/apags/resources/; https://www.apa.org/pi/dis-
ability; https://scn40.org/piac-ema/; https://www.nanonline.org (click About NAN > NAN
Committees > Culture & Diversity Committee). We include more links at www.NavNeuro.com/
book.

are already sold on the idea of becoming a neuropsychologist and simply want step-by-step guidance about how to get there, then feel free to read Part II first. If you do, Part I can serve as a resource for you to fall back on any time you have questions about the field.

Brief Overview

Although Part II provides a detailed description of each step on the path to becoming a neuropsychologist (also see the beginning of Ch. 4), we think that it is worthwhile to provide a brief overview at the outset. This will give you a taste of what is to come. The following is the traditional route to becoming a neuropsychologist (see the Figure).[4]

First, we graduate from college with a bachelor's degree, having completed courses in psychology, biology, and neuroscience, among others. Next, we attend graduate school for 5–6 years, earning a doctorate (a PhD or a PsyD) in clinical psychology. During graduate school, we receive broad training in topics such as statistics, research methods, mental illness, psychological assessment, psychotherapy, cultural diversity, and professional ethics. We also begin our specialty training in neuropsychology through work specifically related to injuries and disorders of the brain and behavior such as learning disorders, autism spectrum disorder, depression, anxiety, schizophrenia, traumatic brain injury, stroke, epilepsy, Alzheimer's disease, Parkinson's disease, and many others. In the final year of graduate school, we complete a psychology internship followed by a 2-year postdoctoral fellowship ("postdoc" for short). The internship is a broad, clinically focused year that rounds out our graduate training in clinical psychology, while the postdoc is our capstone training experience. It is during postdoc that we gain depth of training and truly specialize in neuropsychology. Next, if we want to provide clinical services to our own patients (most neuropsychologists do), we pursue licensure as a psychologist in a particular state. Finally, many neuropsychologists opt to become board certified, which signifies the highest level of clinical accreditation and achievement in our field.

Of course, our learning does not end when we enter the workforce as an independent neuropsychologist. Imagine visiting your doctor, who has been practicing medicine for 40 years, and discovering that they stopped reading the scientific literature the moment they became a licensed MD. All of their knowledge and clinical decisions would be based on research from the pre-1980s era! That would be a scary office visit, indeed. We avoid this type of scenario by spending our entire careers keeping abreast of progress and innovation, reading the scientific literature, attending professional conferences, and participating in didactics.[5] We are lifelong learners, through and through.

---

[4] Where possible, we discuss non-traditional routes throughout the book.

[5] "Didactics" refers to lectures, seminars, brown-bag lunches, clinical case presentations, and other formal educational experiences.

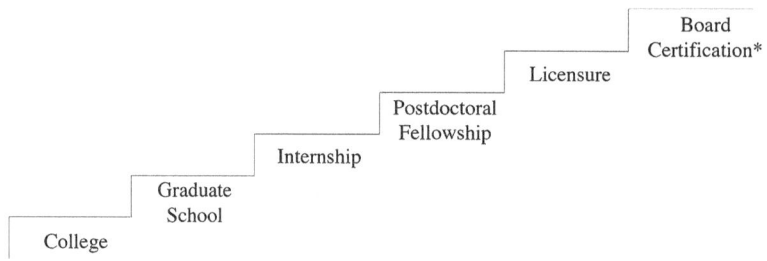

Figure. Steps to becoming a clinical neuropsychologist.
*Board certification is currently optional

Can *I* Become a Neuropsychologist?

After reading the last few paragraphs, you may be thinking, "Neuropsychology sounds right up my alley!" However, if you are like many people, your very next thought will be, "But surely I don't have what it takes to get there." We disagree with this sentiment. There is no "special sauce" required to become a neuropsychologist. You needn't be a savant, a math whiz, or a model of popularity in order to work in this field (we're certainly none of these). We want to assure you that if you care about helping people and are capable of working hard, delaying gratification, and thinking critically, then you already have several of the key ingredients necessary to be a successful neuropsychologist.

Another limiting belief that may prevent students from entering the field of neuropsychology is thinking that they are too "different" from the typical student who pursues this path. This difference may be in the form of cultural factors, economic status, employment/academic background, or other characteristics. However, we mentioned earlier that diversity is essential to ensuring a future for the field, and you may be surprised to know that individual journeys into the discipline are quite diverse and often nonlinear. Take Dr. Steve Correia, currently a board-certified neuropsychologist at Brown University, who left his first occupation as a musician and high school band leader to pursue graduate school in neuropsychology. Another example is one of us (Ryan), who worked for several post-high school years as a "produce boy" and a truck driver before finishing college and ultimately pursuing neuropsychology.

Importantly, you do not need to be born in the US in order to pursue a career in the field. Zanjbeel Mahmood, a current graduate student at UCSD, overcame financial concerns related to attending college while simultaneously navigating her identity as a mixed-race, adult immigrant to the US. She attended community college and held multiple jobs throughout her undergraduate career. Serendipitously, the economic factors that motivated Zanjbeel to seek paid employment while in college led to her introduction to the field of clinical neuropsychology, as she obtained a research assistantship at the Semel Institute of Neuroscience and Human Behavior and transferred to UCLA to earn her bachelor's degree. She is now well on her way to becoming a neuropsychologist. And we will end with the story of Dr. Jean Ikanga. Dr. Ikanga was born and raised in a rural village in the Democratic Republic of the Congo, left his country to pursue training in neuropsychology, and subsequently returned to become the first neuropsychologist in the Congo and one of the few on the continent. He currently holds joint appointments at the University of Kinshasa (Congo) and Emory University.[6]

Coming back to our original point, we think that these stories are important because neuropsychology benefits from people with a wide range of life experiences. So, being "different" can be an asset rather than a liability. Personal characteristics such as being born outside of the US, having work experience in areas other than psychology and medicine, or being raised in a household with diverse belief systems all lead us to see the world differently. These experiences allow neuropsychologists to connect with people and to think about problems in unique and creative ways, which benefits our field.

Caveats

A Note About Geography

Multiple sections of this book are US specific (e.g., pertaining to Veterans Affairs), and much of the content is aimed at an American audience. This is because the US has been the epicenter of many important developments in the field and it is the system with which we (John and Ryan) are most familiar. Additionally, neuropsychology training and practice can look very different in different countries and jurisdictions, and an in-depth discussion of other models is outside the scope of this book. However, we strongly encourage people all over the world to consider pursuing careers in neuropsychology and we believe that much of the content of this book is applicable and useful to international readers. To further this effort, we have created a webpage with resources pertaining to neuropsychology outside of the US: www.NavNeuro.com/global.[7]

---

[6] There are many other neuropsychologists with unique paths from all over the globe. Visit www. NavNeuro.com/MyStory for additional examples.

[7] A list of all website links (and resources) included in the book can be found at www.NavNeuro. com/book.

A Note About Terminology

Although this book is about becoming a neuropsychologist, the term "neuropsychologist" is not unanimously defined and agreed upon. We have taken a conservative approach by defining a clinical neuropsychologist in the US as someone who is eligible for board certification in the field, but we acknowledge that there are many brain-behavior scientists who are not interested in board certification (because it is a clinical credential), yet who can legitimately call themselves neuropsychologists.

Disclaimers

1) The contents of this book do not represent the views of any neuropsychological organization and we do not speak for all neuropsychologists.

2) This book is intended as a beginner's resource and is not to be used as a substitute for formal training.

3) Our step-by-step coverage of the path to becoming a neuropsychologist in Part II is meant to be informative rather than prescriptive. We do not intend to be the gatekeepers for who should or should not be qualified to be called a neuropsychologist. Rather, Part II represents the comprehensive guidance that we would provide to an interested student, and that we wish we had been given when we were students ourselves. This guidance comes from our experience having recently completed our own training, from talking with numerous neuropsychologists, and from background readings and research. If you have a question about your specific training or your eligibility to become a neuropsychologist, we suggest that you speak with your supervisors, review the resources we provide, and contact the relevant board (see Chapter 7).

John A. Bellone
Ryan Van Patten

Acknowledgments

We would like to express our sincere gratitude to each of the following people for reviewing parts of the manuscript, offering guidance throughout the writing process, contributing quotes, and/or mentoring us on our journeys to becoming neuropsychologists:

Kevin Apple, PhD, Associate Dean and Professor at James Madison University

Kira Armstrong, PhD, ABPP-CN, Neuropsychologist in private practice

Peter Arnett, PhD, Neuropsychologist and Professor at Pennsylvania State University

William Barr, PhD, ABPP-CN, Neuropsychologist and Associate Professor at New York University-Langone Health

Timothy Belliveau, PhD, ABPP-CN, Neuropsychologist and Assistant Professor at Yale University

Danielle T. Bello, PhD, ABPP-CN, Neuropsychologist at Cure 4 The Kids Foundation

Raquel Bellone, BA, Social Services Case Worker

Ilex Beltran-Najera, MA, Neuropsychology Doctoral Student at the University of Houston

Adam Brickman, PhD, Neuropsychologist and Professor at Columbia University

Josef Cohen, Undergraduate Student at Columbia University

Meghan Collier, PhD, Neuropsychologist at Rhode Island Cognitive Behavioral Therapy (RICBT)

Steve Correia, PhD, ABPP-CN, Neuropsychologist at Butler Hospital and Associate Professor at Brown University

Munro Cullum, PhD, ABPP-CN, Neuropsychologist in the Peter O'Donnell Brain Institute at the University of Texas Southwestern Medical Center

Jennifer Davis, PhD, ABPP-CN, Neuropsychologist and Associate Professor at Brown University and Rhode Island Hospital

Liselotte de Wit, MS, Neuropsychology Graduate Student at the University of Florida and Researcher at Radboud University in the Netherlands

Dean Delis, PhD, ABPP-CN, Neuropsychologist and Emeritus Professor at UCSD

Peter Dodzik, PsyD, ABPdN, ABN, Board-Certified Clinical Neuropsychologist and Clinical Associate Professor of Neurology at Indiana University

Sean Evans, MD, Neurologist and Associate Clinical Professor at UCSD

Lisa Eyler, PhD, Neuropsychologist and Professor at UCSD

Peter Farenkamp, MA, Organizational Development and Social Transformation

Travis G. Fogel, PhD, ABPP-CN, Neuropsychologist and Assistant Professor at Loma Linda University

Kathleen Fuchs, PhD, ABPP-CN, Neuropsychologist and Associate Professor at the University of Virginia

Charles Gaudet, PhD, Neuropsychology Postdoctoral Fellow at Brown University

Leslie S. Gaynor, MS, Neuropsychology Graduate Student at the University of Florida

Igor Grant, MD, Psychiatrist, Distinguished Professor, and past Chair of Psychiatry at UCSD

Taylor Greif, PhD, Neuropsychology Postdoctoral Fellow at the University of Michigan

Thomas Guilmette, PhD, ABPP-CN, Neuropsychologist, Professor at Providence College, and Adjunct Associate Professor at Brown University

Richard E. Hartman, PhD, Behavioral Neuroscientist and Professor at Loma Linda University

Bruce Hermann, PhD, ABPP-CN, Neuropsychologist and Professor Emeritus at the University of Wisconsin School of Medicine and Public Health

Julie Hook, PhD, MBA, ABPP-CN, Neuropsychologist, Research Associate Professor at Northwestern University, and Product Manager of the NIH Toolbox®

Jean Ikanga, PhD, Staff Scientist at Emory University School of Medicine and Assistant Professor at the University of Kinshasa's School of Medicine.

Joel Kamper, PhD, ABPP-CN, Neuropsychologist at the James A. Haley Veterans Hospital

David A. S. Kaufman, PhD, ABPP-CN, Neuropsychologist and Clinical Associate Professor at Saint Louis University

Christine Koterba, PhD, ABPP-CN, Neuropsychologist at Nationwide Children's Hospital and Clinical Assistant Professor at Ohio State University

Laura Lacritz, PhD, ABPP-CN, Director of Neuropsychology at the University of Texas Southwestern

Margaret Lanca, PhD, Assistant Professor of Psychology at Harvard Medical School and Director of Adult Neuropsychology and Psychological Testing at the Cambridge Health Alliance

Robert D. Latzman, PhD, Psychologist and Associate Professor at Georgia State University

Zanjbeel Mahmood, MS, Neuropsychology Graduate Student at UCSD

Toni Maraldo, PhD, Licensed Psychologist, Truman VA Hospital

Karen Miller, PhD, Neuropsychologist and Professor at UCLA

Andrea Mustafa, MA, Neuropsychology Graduate Student at San Diego State University

Nathaniel Nelson, PhD, ABPP-CN, Neuropsychologist and Associate Professor at the University of St. Thomas, Graduate School of Professional Psychology

Tanya Nguyen, PhD, Neuropsychologist and Assistant Professor at UCSD

Nancy Nussbaum, PhD, ABPP-CN, Neuropsychologist and Associate Professor at Dell Children's Medical Center, University of Texas at Austin

Alexis Olson, PhD, Neuropsychologist in private practice

Adam Parks, PhD, ABPP-CN, Neuropsychologist and Assistant Professor at the University of Kansas Medical Center

Michael Parsons, PhD, ABPP-CN, Neuropsychologist and Assistant Professor at Harvard Medical School and Massachusetts General Hospital

Suzanne Penna, PhD, ABPP-CN, Neuropsychologist and Associate Professor at Emory University

Robin Peterson, PhD, ABPP-CN, Neuropsychologist and Assistant Professor at the University of Colorado

Neil Pliskin, PhD, ABPP-CN, Neuropsychologist and Professor at the University of Illinois at Chicago

Karen Postal, PhD, ABPP-CN, Neuropsychologist in private practice and instructor at Harvard Medical School

Beth Slomine, PhD, ABPP-CN, Neuropsychologist and Professor at Johns Hopkins University and the Kennedy Krieger Institute

Glenn Smith PhD, ABPP-CN, Neuropsychologist and Professor and Chair of Clinical and Health Psychology at the University of Florida

Megan Spencer, PhD, Neuropsychologist and Assistant professor at Brown University and the Providence VA Medical Center

Anthony Stringer, PhD, ABPP-CN, Neuropsychologist and Professor at Emory University

April Thames, PhD, Neuropsychologist and Associate Professor at the University of Southern California

Ryan Townley, MD, Neurologist and Assistant Professor at the University of Kansas

Christine Trask, PhD, ABPP-CN, Neuropsychologist at Lifespan and Clinical Assistant Professor at Brown University

Geoffrey Tremont, PhD, ABPP-CN, Neuropsychologist and Associate Professor at Brown University and Rhode Island Hospital

Elizabeth W. Twamley, PhD, Professor at UCSD and Director of the Clinical Research Unit, Center of Excellence for Stress and Mental Health (CESAMH), VA San Diego Healthcare System

Adam J. Woods, PhD, Psychologist, Cognitive Neuroscientist, and Associate Professor at the University of Florida

Jeff Wozniak, PhD, LP, Neuropsychologist and Professor at the University of Minnesota

John D. Wright, PhD, ABPP-CN, Neuropsychologist at Mercy Hospital in Saint Louis

Keith Owen Yeates, PhD, ABPP-CN, Neuropsychologist and Professor at the University of Calgary

We would also like to thank all of the students who asked us about becoming a neuropsychologist. They inspired us to write the book.

Contents

About the Authors

John A. Bellone John Bellone earned his PhD in Clinical Psychology (neuropsychology/neuroscience concentration) at Loma Linda University, including completing a 1-year practicum at UCLA and a 1-year clinical internship in neuropsychology at Yale University School of Medicine. He completed a 2-year neuropsychology postdoctoral fellowship at the Alpert Medical School of Brown University. Dr. Bellone currently works as a clinical neuropsychologist in a group practice in Southern California, working in both outpatient and inpatient rehabilitation settings. He is a member of multiple neuropsychological organizations and has recently obtained board certification in clinical neuropsychology through the American Board of Professional Psychology (ABPP). Additionally, he is the backup neuropsychology consultant for the Anaheim Ducks NHL team. Although his clinical interests are broad, he is particularly passionate about reducing risk for cognitive decline and improving overall health through lifestyle modification. He frequently lectures on the power that exercise, healthy diet, quality sleep, psychological well-being, and staying cognitively and socially active have on maintaining/improving cognitive functioning both now and as we age.

Ryan Van Patten Ryan Van Patten earned his PhD in clinical psychology (neuropsychology concentration) at Saint Louis University, including completing a 1-year practicum at Washington University in St. Louis and a 1-year clinical internship in neuropsychology at the Alpert Medical School of Brown University. He completed a 2-year postdoctoral fellowship in neuro-psychology at UCSD. Dr. Van Patten is currently working in neuropsychological research and patient care at Massachusetts General Hospital, Spaulding Rehabilitation Hospital, and Harvard Medical School in Boston, MA. His research and clinical interests include cognitive aging, neurodegenerative diseases, traumatic brain injury, stroke, severe mental illness, cognitive interventions, and technology in neuropsy-chology. Dr. Van Patten is also interested in the educa-tion, mentoring, and teaching of neuropsychological trainees.

Drs. Bellone and Van Patten are the co-creators and co-hosts of the podcast, Navigating Neuropsychology, where they provide evidence-based educational content to students, trainees, and professionals by interviewing experts and discussing a variety of topics within neuro-psychology. Visit www.NavNeuro.com for more information.

Acronym List

AACN	American Academy of Clinical Neuropsychology
AAPI	APPIC Application for Psychology Internships
ABCN	American Board of Clinical Neuropsychology
ABN	American Board of Professional Neuropsychology
ABPP	American Board of Professional Psychology
ADHD	Attention Deficit/Hyperactivity Disorder
AIDS	Acquired Immunodeficiency Syndrome
AMCs	Academic Medical Centers
ANST	The Association of Neuropsychology Students and Trainees
AP	Advanced Placement
APA	American Psychological Association
APAGS	American Psychological Association of Graduate Students
APPCN	Association of Postdoctoral Programs in Clinical Neuropsychology
APPIC	Association of Psychology Postdoctoral and Internship Centers
ASD	Autism Spectrum Disorder
ASPPB	Association of State and Provincial Psychology Boards
AVM	Arteriovenous Malformation
BRAIN	Be Ready for ABPP in Neuropsychology
BRAIN	Brain Research Through Advancing Innovative Neurotechnologies
CBT	Cognitive Behavioral Therapy
CDA	VA Career Development Award
CECP	Committee on Early Career Psychologists
CFP	Certified Financial Planner
CLEP	College-Level Examination Program
CNS	Clinical Neuropsychology Synarchy
CPA	Canadian Psychological Association
CPCRT	Certification in the Practice of Cognitive Rehabilitation Therapy
CT	Computed Tomography
CV	Curriculum Vitae
DCT	Director of Clinical Training
ECNPC	Early Career Neuropsychologist Committee

ECT	Electroconvulsive Therapy
EEG	Electroencephalogram
EPPP	Examination for Professional Practice in Psychology
ETS	Educational Testing Service
FASD	Fetal Alcohol Spectrum Disorder
fMRI	Functional Magnetic Resonance Imaging
GPA	Grade Point Average
GRE	Graduate Record Examination
HABIT	Healthy Action to Benefit Independence and Thinking
HC	Houston Conference
HCG	Houston Conference Guidelines
HIV	Human Immunodeficiency Virus
ICU	Intensive Care Unit
IED	Improvised Explosive Device
IEP	Individualized Education Program
INS	International Neuropsychological Society
LBD	Lewy Body Dementia
MAMBIT	Mental Abilities as Measured by Intelligence Tests
MCI	Mild Cognitive Impairment
MD	Doctor of Medicine
MRI	Magnetic Resonance Imaging
NAN	National Association of Neuropsychology
NANSPRC	NAN Student & Post-Doctoral Resident Committee
NASA	National Aeronautics and Space Administration
NIH	National Institute of Health
NMS	National Matching Services, Inc.
NOSI	Neuropsychology Outcome Satisfaction Initiative
OEF	Operation Enduring Freedom
OIF	Operation Iraqi Freedom
OR	Operating Room
OT	Occupational Therapy
PCP	Primary Care Physician
PET	Positron Emission Tomography
PhD	Doctor of Philosophy
PI	Principal Investigator
PMVS	Post-Match Vacancy Service
PSLF	Public Service Loan Forgiveness
PsyD	Doctor of Psychology
PSYPACT	Psychology Interjurisdictional Compact
PTSD	Posttraumatic Stress Disorder
RA	Research Assistant
ROTC	Reserve Officers' Training Corps
SAC	Student Affairs Committee
SCN	Society for Clinical Neuropsychology
SLC	Student Liaison Committee

SLP	Speech-Language Pathologist
TA	Teaching Assistant
TBI	Traumatic Brain Injury
TCN	The Clinical Neuropsychologist
tDCS	Transcranial Direct Current Stimulation
TMS	Transcranial Magnetic Stimulation
VA	Veterans Affairs

Part I
The Brain-Behavior Landscape

Chapter 1

What Is Neuropsychology?

Welcome! I'm Dr. Bellone. Please come in and have a seat. Many people are not exactly sure who I am or what this is all about, so I'll give you a quick overview. I'm a neuropsychologist. That's a fancy way of saying that I'm a psychologist with specialty training in brain function and dysfunction. You are here for a neuropsychological evaluation, which takes place in two parts. The first part involves me asking you questions in order to learn important information… questions such as the nature of your cognitive problems, your medical history, and your social background. This will help me get to know you better so that I can tailor my opinions and recommendations to you. The second part involves objectively testing your thinking skills such as memory, attention, and problem solving using paper-pencil and computer-based tests. The purpose of this is to help me understand your cognitive strengths and weaknesses; we are all better at some tasks than others and if I know your unique profile, I can make the most accurate diagnosis and suggest ways to improve your health and quality of life.

The entire evaluation typically takes about three to four hours. When you leave, I will score the tests and compare your performance to other people of a similar age and education level. Then I will write up what I've learned about you in a medical report so that your doctors can use it to provide you with the best possible care. Finally, I like to sit down with people on a separate day, once everything else is completed, so that I can share my findings, provide recommendations, and give you the opportunity to ask any questions that you might have. How does this sound? Do you have any questions so far?

This is how I (John) typically introduce the neuropsychological evaluation to my patients and I thought that it would be fitting to share it with you right up front. Although this is a relatively standard approach, there are multiple flavors or varieties to the basic template, and the specific language used is largely contingent upon the setting, the patient, and the neuropsychologist themself.

****

© Springer Nature Switzerland AG 2021
J. A. Bellone, R. Van Patten, *Becoming a Neuropsychologist*,
https://doi.org/10.1007/978-3-030-63174-1_1

Neuropsychology is a fascinating area of study (we're not biased at all!). It can be loosely defined as the scientific study of how the brain produces behavior and how behavior is altered when something atypical happens to the brain (the "brain-behavior relationship").[1,2] That is, our cognitive abilities, emotional states, personality traits, and overt actions can all be traced back to brain activity that occurs in complex, interconnected networks (Kolb and Whishaw 2015). In other words, these few pounds of jelly between our ears are responsible for *everything* that we do, think, and feel. If something changes the structure and functioning of our brain (e.g., a severe traumatic brain injury), or if the brain did not develop normally to begin with (e.g., megalencephaly; Mirzaa and Poduri 2014), we see alterations in thinking and actions. For example, imagine that a man has a stroke that cuts off blood supply to his left frontal lobe, adjacent to the Sylvian fissure. It is likely that this injury would leave him with significant difficulties in expressing his thoughts verbally. In this case, speech is the behavior that is affected.

Throughout the rest of the chapter, we will cover the purpose of neuropsychology, the types of patients seen, the cognitive skills assessed, and the differences relative to related professions.[3] However, before we get into these weeds, we will briefly discuss past approaches for examining brain-behavior relationships because this rich history laid the foundation for the field of neuropsychology as we know it today.

Roots of Neuropsychology/Historical Context

One thing that I try to pass on to trainees is not to be too short-sighted. We should all appreciate the giants on whose shoulders we're standing, and the history of our field.

– Lisa Eyler, PhD

People have been interested in the cause of thinking and behavior for thousands of years. The earliest available evidence for this comes from ancient skulls with burr holes (Fig. 1.1). It is thought that these holes were created to release evil spirits or to facilitate the introduction of benevolent spirits, and as a primitive surgical intervention to treat injuries (any volunteers??). The procedure is called "trepanning" or "trepanation" and was widespread, having been developed independently through-

---

[1] Scientists use the term "behavior" in two different ways. Here we are using it broadly so that it includes not only overt actions but also thought processes and affective states. The other meaning of "behavior" (e.g., "behavioral observations") is narrower and refers specifically to observable actions, not cognition and emotion.

[2] There are slight variations in the definition of "neuropsychology." For example, see The American Academy of Clinical Neuropsychology practice guidelines (Board of Directors 2007) and the National Academy of Neuropsychology's position paper (Barth et al. 2003).

[3] Also, listen to NavNeuro episode 2, *Neuropsychology for Non-Neuropsychologists*, for an overview (www.NavNeuro.com/02).

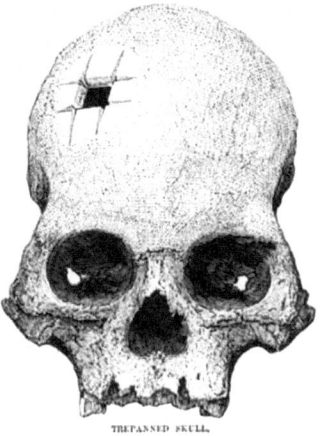

TREPANNED SKULL.

Fig. 1.1 A skull with a burr hole

out Europe, China, and the Americas, with 5–10% of all discovered Neolithic skulls (circa 6500 BCE and earlier) showing evidence of the practice (Faria 2015).

The earliest known written mention of the brain comes from a document made by an Egyptian physician around 1600 BCE (although it may be a copy of a manuscript dating back to 3000 BCE) called the *Edwin Smith Surgical Papyrus* (Kamp et al. 2012). The document describes 27 battlefield head injury cases, including the nature of the injury, examination, and diagnosis. At the time, the brain was generally not recognized as an important organ, as evidenced by the Egyptian practice of throwing it away when preparing a corpse for embalming; in contrast, the heart was thought to be the center of intelligence. Other ancient civilizations such as the Mesopotamian, Hindu, and Hellenic peoples held similar views about the heart being central to thought and emotion (Faria 2015).[4] Even many esteemed scholars of antiquity did not understand the importance of the brain. For example, Aristotle believed that the brain's purpose was to cool the blood and heart (Aristotle, ca. 350 BCE/ 1912).

Although early prevailing views of the brain's function were based on philosophical and religious beliefs rather than careful scientific examination, there were some who used anatomical study as early as the fifth and sixth centuries BCE to begin understanding the importance of the brain for higher-level abilities (Konstantine and Peter 2015). Anatomical study became more and more common over the next few centuries; for example, Galen dissected the brains of animals in the second century CE. By ~1000 CE, Middle Eastern physicians were regularly performing surgical procedures for brain-related issues (e.g., Al-Zahrawi; Al-Rodhan and Fox 1986) and presenting detailed accounts of brain injuries and treatments (e.g., Ibn Sina; Aciduman et al. 2009). The first European anatomy textbooks began emerging in the

---

[4]Remnants of this sentiment are evident in various idioms, such as "to know by heart" or "you will always be in my heart."

thirteenth and fourteenth centuries (e.g., *Anathomia Corporis Humani*), and anatomical inquiry blossomed throughout the Renaissance (Nanda et al. 2016).

The study of brain-related problems through anatomy continues today in the field of neuropathology. Modern neuropathology began in the late 1800s with the discoveries of Santiago Ramón y Cajal, Camillo Golgi, and others. A neuropathological examination typically involves dissecting a human brain after the person dies, both visually inspecting the gross anatomy and cutting small pieces in order to examine neurons and glial cells under a microscope. Various staining methods can be applied to the tissue to visualize different components or pathology. For example, in 1906 Alois Alzheimer used these techniques to view the amyloid plaques and neurofibrillary tangles that would become known as hallmarks of Alzheimer's disease. Eventually, in vivo biopsies became common – this is where pathologists study a resected portion of a living person's brain in order to assess for various disease processes such as whether or not a mass of cells is cancerous.

Because it is obviously not easy or without risks to open up the skull and examine what is going on inside, people have searched for other methods of investigating what is happening up there in a living person. One popular early method, "phrenology," originated in 1796 based on the work of German physician Franz Joseph Gall and was popularized by Johann Spurzheim. The technique involved performing a "skull reading" in which the phrenologist would assess the contours of the outer surface of the skull (especially the indentations and protuberances), and then associate the measurements with cognitive and personality abilities/traits. For example, a phrenologist might conclude that their patient was particularly witty, benevolent, or combative based on feeling the outside of the skull (see Fig. 1.2). Although recognized as pseudoscience by the 1840s (the topography of the skull does not correlate with brain functioning), it helped promote the idea that different regions of the brain relate to different mental abilities, which is often referred to as "localization" or "functional specialization" (see Kolb and Whishaw 2015, for a more thorough discussion). This

Fig. 1.2 A phrenological map of characteristics and abilities

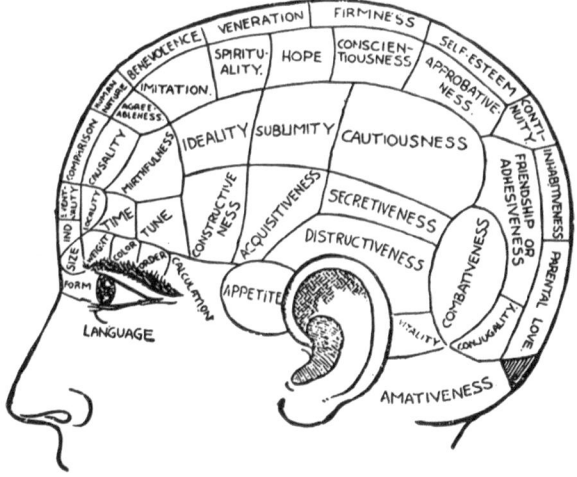

concept is perhaps most concisely summed up by Norman Geschwind's quote, "every behavior has an anatomy" (Geschwind 1975, p. 3).

During and after the phrenology craze, functional specialization became more widely recognized. Perhaps the most famous example illustrating this principle is that of Phineas Gage. In 1848, Gage was a construction foreman preparing a road-bed for a railroad when an accidental explosion sent a three-and-a-half foot, 13-pound tamping iron through his left jaw and out of his frontal lobe (Harlow 1848). Surprisingly, he survived the injury and ultimately made an incredible recovery, but his physician noted that his personality had been drastically altered. The man people knew to be reserved and considerate was now impulsive and profane; the physician noted that he "was no longer Gage" (Harlow 1868). Although there is debate about whether the degree or persistence of Gage's personality change was exaggerated (Macmillan 2000), it is clear that this case helped people begin to understand that personality and social behavior can be modified by injury to specific brain regions.

Given the growing confidence that it is possible to acquire knowledge pertaining to what is happening in the brain by observing a person's behavior, many physicians began correlating neuroanatomy and neuropathology with clinical observations in a systematic way. This led to what has been referred to as the "golden age of cerebral localization" (Benton 2000, p. 223). For example, Paul Broca performed autopsies on his patients and noticed that people who had difficulty speaking fluently often had damage to a specific part of their brain (Broca 1865). Thus, if a patient presented with an expressive language deficit, the physician could infer that it was likely caused by damage in the left posterior inferior frontal gyrus. Shortly after Broca's discovery,[5] Carl Wernicke observed that a different type of language problem involving poor comprehension and fluent but incoherent speech was linked to damage to the posterior superior temporal gyrus (Wernicke 1874; see Fig. 1.3). A more recent case study of brain-behavior relationships involves a patient called "H.M.," who had both of his medial temporal lobes surgically removed in 1953 as a treatment for intractable epilepsy.[6] Although the procedure partially controlled his seizures, it robbed him of his ability to form memories of new events (Milner et al. 1968). So, within minutes of being presented with information, the material would be completely forgotten (think of the 2000 movie *Memento* or the 2004 movie *50 First Dates*). This case helped clinicians better understand how learning and memory work – i.e., structures in the brain known as the hippocampus and parahippocampal regions play an integral role in memory formation. Many more of these

---

[5] Another French neurologist (Marc Dax) actually made this discovery ~30 years earlier but it was not publicized until after Broca's famous paper, so the finding is generally associated with Broca instead of Dax (Drouin and Péréon 2019). It is thought that Broca's discovery was independent of Dax's work.

[6] It is meaningful that both medial temporal lobes were removed because we have one in each of our two cerebral hemispheres (right and left) and each can partially compensate for damage to the other.

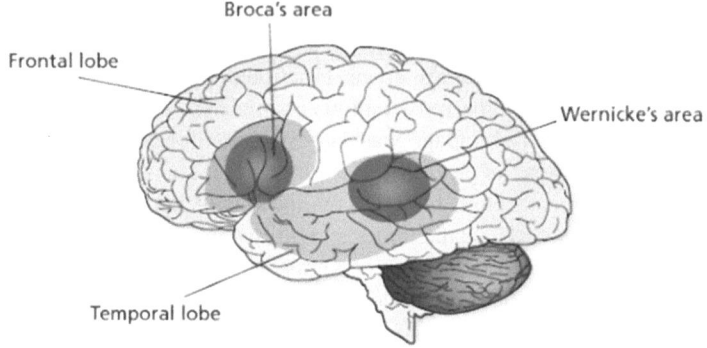

Fig. 1.3 Language areas in the brain

types of cases emerged, with additional major anatomical regions being related to specific behaviors.[7,8]

Early Days of Neuropsychology

Neuropsychology emerged in part from the flurry of localization insights in humans and animals. It is impossible to pinpoint exactly when the field was born, but the term "neuropsychology" was first used in a 1913 address by William Osler, who viewed the field as the study of the impact of brain injury, disease, and/or mental illness on "higher functions" (Bruce 1985).[9] Because there was no easy way to peer inside the brain of a living person at the time, the field initially focused on lesion localization – i.e., estimating the location of brain injury based on an assessment of cognitive abilities. As you would expect, there was much need for this during and after World War I (WWI; 1911–1916) and World War II (WWII; 1939–1945), when soldiers were wounded by bullets, shrapnel, and other weapons. Additionally, there was a need for evaluations of military personnel prior to battlefield engagement. In this vein, in 1917, President of the American Psychological Association Robert Yerkes persuaded the US Army that conducting intelligence testing on military

---

[7] We encourage readers to check out books by the late Oliver Sacks, a behavioral neurologist who describes interesting brain-behavior relationships in an entertaining way (e.g., *The Man Who Mistook His Wife for a Hat*; and *An Anthropologist on Mars*).

[8] This was only a very brief overview of the history of neuroscience and does not include important theories such as connectionism, equipotentiality, and hierarchical organization. For more thorough accounts, see the following books, both by Stanley Finger: (1) *Origins of Neuroscience: A History of Explorations Into Brain Function*, and (2) *Minds Behind the Brain: A History of the Pioneers and Their Discoveries*.

[9] For more information about the origins of the term "neuropsychology," including its use by Karl Lashley and Kurt Goldstein, see Finger (1994).

recruits would help them select the best soldiers and assign them to appropriate tasks. Along with Henry Goddard, Lewis Terman, and others, Yerkes developed the Alpha and Beta tests for those who were literate and those who were illiterate, respectively (there was also a modified test for those who failed the Beta version). To create these measures, they capitalized on tests that had been recently developed such as the Binet-Simon test and the Stanford-Binet Intelligence Scales. They started the assessment program in 1918 and tested about 1.7 million recruits, making this the largest cognitive assessment project of the time (Shephard 2015). There is a complex and rich history behind intelligence testing and the evolution of psychology in general, but it is beyond the scope of this book.[10]

A scientifically grounded understanding of intelligence played an important role in the genesis and development of neuropsychology by increasing funding and jobs in psychology, particularly in the area of cognitive assessment (Flanagan and Harrison 2018).[11] Importantly, the newfound popularity of intelligence testing also acted as an incubator for the development of statistical methods and psychometrics (the study of measuring cognitive abilities), which are integral to the theory and practice of neuropsychology.

By the time of WWII, a psychologist named David Wechsler had just created the precursor to his intelligence scales, which are still widely used by neuropsychologists today (Rabin et al. 2016).[12] The war and its aftermath provided no shortage of patients, and many brain-behavior enthusiasts offered their knowledge and skills to help these Veterans. A secondary benefit of this movement was that it advanced our understanding of brain functioning broadly and traumatic brain injuries specifically. One particularly prominent figure in the post-WWII era was Alexander Luria, a Russian neuropsychologist who was integral to the development of assessment practices (Kostyanaya and Rossouw 2013). Among his many accomplishments was the development of various cognitive and sensory tests, as well as contributions to the systematic description of functional neural systems (Luria 1966). His work provided the theoretical basis that led to the development of the Luria-Nebraska Neuropsychological Battery, a popular assessment method in the late twentieth century (Golden et al. 1979). Other prominent pioneers of neuropsychological assessment in the WWII/post-WWII era include Ward Halstead and Ralph Reitan, who collaborated on the Halstead-Reitan battery. This family of tests gained widespread popularity among clinicians and researchers who were looking for a comprehensive, data-driven method for determining the presence of neurocognitive impairment. Importantly, Luria, Halstead, and Reitan were not alone; the drive to continue

---

[10] If you are interested in learning about other key players, look up the work of Charles Spearman, Raymond Cattell, John Horn, and John Carroll.

[11] There were glimmers of systematic neuropsychological assessment prior to this intelligence testing period; See Benton (2000) for details.

[12] The Wechsler-Bellevue Intelligence Scale came out in 1939. The Wechsler Intelligence Scale for Children (WISC) was published in 1949 and was made up of many of the original measures. The Wechsler Adult Intelligence Scale (WAIS) came out in 1955. As of the time of writing, the WISC is currently in its fifth edition and the WAIS is in its fourth edition.

generating and refining our theories and methods has led to major contributions by countless other men and women.[13]

****

Although the principles and underlying science of neuropsychology are centuries old, the field as a coherent discipline is quite young (Grote et al. 2016). For example, the first neuropsychology journal, *Neuropsychologia*, was established in 1963. Our oldest professional society, the International Neuropsychological Society (INS) was founded in 1967. The first two textbooks specific to the field came out in 1976 and 1980.[14] The first formal training programs were established in the 1970s, with specialty board certification developing in 1981 and formal training guidelines being published in 1987 (INS-Division 40 Task Force 1987) and 1998 (i.e., Hannay et al. 1998). Consequently, we believe that the field is both built on a solid foundation of philosophical and scientific inquiry, and is still in its early adolescent phase of development. This makes neuropsychology ripe for the novel ideas and creative thinking that is characteristic of newcomers such as yourself!

Modern-Day Neuropsychology

Purpose of Clinical Evaluations

The practice of neuropsychology today looks very different than it did in the early years. One of the stronger pressures driving the shift in the field was the advent of widely available neuroimaging tools in the 1980s. The ability to noninvasively scan brain tissue with computed tomography (CT), magnetic resonance imaging (MRI), and other techniques reduced the value of neuropsychological testing for the purpose of lesion localization. With this technology, physicians were able to send a patient through a scanner and visualize the size, type, and location of a lesion.

Some neuropsychologists initially feared that neuroimaging would put them out of a job, but we are a resilient bunch! Instead of closing up shop, clinicians and researchers began expanding their roles and areas of expertise. The emphasis of our evaluations shifted from lesion localization to (1) clinical diagnosis, (2) characterizing cognitive strengths and weaknesses, (3) determining appropriate interventions and services, and (4) cognitive training and rehabilitation. Let's talk more about each of these tasks performed by neuropsychologists.

---

[13] Examples of scientist-clinicians who made outstanding contributions to neuropsychology include Arthur Benton, Elizabeth Warrington, Hans-Lukas Teuber, Brenda Milner, Manfred Meier, and Edith Kaplan.

[14] *Neuropsychological Assessment*, by Lezak et al., was first published in 1976; *Fundamentals of Human Neuropsychology*, by Kolb and Wishaw, was first published in 1980.

1. Most of today's neuropsychologists are expert diagnosticians, meaning that we are trained in identifying and labeling various neurological and psychiatric disorders. For example, an older woman presents to her primary care provider with complaints of forgetfulness and the physician refers her to a neuropsychologist to determine the cause of the problem. It might be that she is showing the first signs of Alzheimer's disease, it may be a different type of neurodegeneration (e.g., Lewy body pathology, frontotemporal lobar degeneration), it could be related to a history of heavy alcohol use, she may have depression, or it might just be normal age-related cognitive decline. By reviewing medical records, completing a thorough clinical interview, and conducting comprehensive cognitive testing, we can often parse out the likely etiology (the underlying cause) of the symptoms. We are able to do this because certain diseases and injuries have signature patterns, or "fingerprints," that stand out in comparison to other conditions. For example, a person with Lewy body dementia typically presents with well-formed visual hallucinations, motor symptoms, sleep disturbances, fluctuating cognitive abilities, and poor visuospatial abilities. This cluster of symptoms and performance on testing is unique relative to the typical Alzheimer's disease presentation (i.e., early-stage impairments in memory for recent events).

2. In addition to determining diagnosis and etiology, a major purpose of neuropsychological testing is to characterize cognitive strengths and weaknesses. Everyone has a unique set of mental aptitudes. For example, some people are adept at visual learning but are below average with regard to verbal abilities; others rapidly process information, but have trouble with reasoning and problem-solving. By administering a variety of tests that tap into different types of skills (see below for a description of cognitive "domains"), we get a picture of our patient's individual profile, from impairments to above-average abilities.

3. It is important to know a person's diagnosis, etiology, and strengths and limitations so that we can (a) link their abilities to real-world functioning, (b) estimate their risk for poor outcomes in response to certain medical procedures,[15] (c) select the most appropriate interventions, (d) set goals, (e) direct services, and (f) help them plan for the future. For example, a parent brings their young son to you (a neuropsychologist) because they suspect that he has dyslexia – a type of learning disability in reading. Based on your evaluation, you confirm that he has a reading disability and you also diagnose him with attention-deficit/hyperactivity disorder (ADHD). You carefully construct a list of personalized recommendations based on these disorders and the child's unique situation. Your recommendations include (a) behavior modification for hyperactivity, (b) a referral to a psychiatrist to assess the appropriateness of a stimulant medication, (c) a request for an individualized education program (IEP) to track the boy's educational progress and recommend special accommodations (e.g., 50% extra

---

[15] For example, surgery to resect seizure-producing tissue in someone with epilepsy or deep brain stimulation (DBS) surgery for a patient with Parkinson's disease.

time on tests), and (d) a referral to a tutor who is experienced in educating children with reading disabilities. Based on the severity of the impairment and the cognitive profile, you help the parent understand how the boy is likely to progress through the next few years of school and which services might be needed down the line (e.g., help with social skills and/or employment accommodations).

4. A more recent application of our services – one that is progressing rapidly – is providing cognitive training in order to restore or (more commonly) compensate for cognitive deficits. With regard to the latter, we can use the person's cognitive strengths to help them work around their limitations. For example, there are different types of memory and most people with typical memory disorders have trouble with explicit or declarative memory, where they cannot learn and retain facts and knowledge in their daily lives. Cognitive training for such a person will likely involve capitalizing on intact "implicit" or "procedural" memory.[16] Our patient with an explicit memory disorder may be forgetting novel information such as her plan to call her sister. To take advantage of her implicit memory, we might suggest that she write herself a note the day before she wants to call her sister and stick it to the coffee maker. She knows that she will be making her coffee the next morning (it's an ingrained habit), so even if she forgets to call her sister, she will be reminded of it when she encounters the note. This technique is called "linking" because it involves identifying a behavior that we know will take place (making coffee) and creating a link (the note) between that behavior and the desired behavior (the phone call). This is just one of many examples of cognitive training techniques that have been shown time and time again to improve people's everyday functioning and quality of life (e.g., Greenaway et al. 2013; Huckans et al. 2013; Twamley et al. 2012, 2014).

****

In addition to clinical diagnosis, characterizing strengths and weaknesses, recommending interventions, planning for the future, and conducting cognitive training, neuropsychologists still occasionally perform "lesion localization" evaluations, albeit in a limited set of contexts. For example, presurgical evaluations of epilepsy patients often involve attempting to identify which brain hemisphere (right or left) is more dominant for language and verbal memory so that the surgeon will have an idea about whether removing parts of the brain will leave the patient aphasic (without language) and/or amnestic (without memory; consider the case of H.M., above). Although neuroimaging techniques such as CT and MRI are useful in these circumstances, neuropsychological testing provides unique clinical information that cannot be collected via brain scanning.

---

[16] Implicit memory involves learning habits such as riding a bike, playing golf, or tying your shoes.

Neuroimaging and Neuropsychology

As two classes of tools, neuroimaging and cognitive testing are far from mutually exclusive; in fact, they are complementary. To use an automotive analogy, imagine that you are buying a new car. Structural neuroimaging is similar to examining the car's engine to ensure that everything is intact and in the right place, while cognitive testing is comparable to taking it out for a test drive to ensure that it is working properly. Both of these steps are critical, as you want to know what is under the hood and what happens when you put your foot on the brake pedal.

> One of the origins of neuropsychology was in localizing brain injury. Today, we have neuroimaging techniques that are far better at doing that. But what no imaging technique can do is tell you about what's most important about the brain and that is its output. Output has to do with cognition and behaviors and emotions, and this is where neuropsychology and psychology should be focused.
>
> – Igor Grant, MD

The story behind neuroimaging and cognitive testing goes deeper. It turns out that the search for structure-function relationships (i.e., the localization theory) that was all the rage back in the nineteenth and twentieth centuries is incomplete and has been challenged in recent years.[17] Many studies have shown that cognitive abilities are much more distributed than we had originally thought (Bressler and Menon 2010). In other words, cognitive functions arise from activity in *neural networks,* which span large swaths of interconnected brain tissue, rather than arising from a single discrete region. Consequently, because no part of the brain works in isolation, we should not make statements such as, "memory is in the hippocampus" or "emotion is in the amygdala." Instead, we can make statements such as, "memory is subserved in part by hippocampal circuits" and "emotion regulation is partially mediated by prefrontal-subcortical networks." This may sound overly pedantic and esoteric, but the subtle difference in the above wording is incredibly important, as the network-based phrasing is far more accurate than the localization-based terminology with respect to the organization and function of the brain (see Kolb and Whishaw 2015, for more detail).

Neuropsychology Training

You may be wondering where the "psychology" part of "neuropsychology" comes in, especially because the field has many similarities to neuroscience (the scientific study of the brain and nervous system) and neurology (described below). The distinction between psychology, neuropsychology, and neuroscience/neurology mostly has to do with the knowledge we have, patients we see, and the tools we use. We will briefly cover the first of the three (knowledge) here and the second

---

[17] There were some scientists that challenged it in the past (see Finger 1994), but the push-back did not gain significant traction until recently.

Fig. 1.4 Neuropsychology's location in a simple, non-exhaustive taxonomy of psychological disciplines

(patients) in the following subsection. You can learn all about the tools (i.e., specific tests) by reading the excellent texts by Lezak et al. (2012), and Sherman, Tan, & Hrabok, 2020.

Neuropsychologists are trained as psychologists. This means that we attend graduate school and obtain a doctorate in psychology (either PhD or PsyD; see Ch. 5).[18] We study human development and behavior across different areas of psychology (e.g., social, personality, behavioral, cognitive, clinical). We receive training in psychological therapy ("psychotherapy") and other interventions.[19,20] However, our specialized cognitive assessment skills and in-depth knowledge of brain-behavior relationships are what make us a truly unique discipline relative to the dozens of subspecialties in psychology (see Fig. 1.4).[21]

Similar to many fields within psychology and medicine, neuropsychology has both research and clinical arms. The purpose of the research portion is to generate new knowledge and test interventions through scientific investigation, while *clinical* neuropsychology applies that knowledge to working directly with patients (i.e., "bench to bedside"). Moreover, science and practice support and reinforce each other, where research provides the knowledge base upon which clinical decisions are made, and applied work leads to better research questions (see Fig. 1.5). Many

---

[18] Note that this is how it is done in North America. Also, there are some exceptions that we will discuss in Part II.

[19] In fact, clinical neuropsychologists use psychotherapy skills throughout their careers. Some maintain a psychotherapy practice in addition to assessment; however, even if they do not see patients for therapy, those skills still come in handy during neuropsychological evaluations.

[20] Master's-level practitioners can also engage in therapy, with appropriate credentials and license; however, they are not considered "psychologists," which is a protected term in most states/provinces.

[21] The American Psychological Association currently recognizes 54 distinct disciplines under the umbrella of psychology.

Fig. 1.5 The bidirectional relationship between science and clinical practice

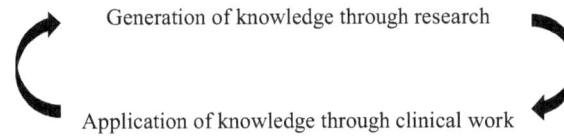

Generation of knowledge through research

Application of knowledge through clinical work

neuropsychologists engage in both research and clinical work, although some focus their entire careers on one or the other.

****

Although the field of neuropsychology has expanded rapidly since its inception, there are still fewer neuropsychologists than is typical in many other professions, so it is considered a niche specialty. Specifically, the best current estimate of the number of neuropsychologists in the US is at least 5765 (C. Morrison, PhD, ABPP-CN, March 2, 2020 personal communication). Of these, 1159 are board-certified through the American Board of Professional Psychology (Armstrong et al. 2019), with approximately 60–80 additional professionals becoming board certified each year (K. Fuchs, PhD, ABPP-CN, August 2, 2020, personal communication).

Clinical Populations Served

Because neuropsychologists specialize in assessing cognitive abilities, any disease, injury, developmental abnormality, or other factor that could affect cognition falls under our purview. One important distinction is that most neuropsychologists consider themselves to be either pediatric (child) neuropsychologists or adult/geriatric neuropsychologists. Similar to physicians, we draw an imaginary line in the sand between age 17 and 18 and most of us spend our time working with one group or the other.[22] There are some lifespan neuropsychologists out there – people who work with both children and adults – but these practitioners are in the minority.

Examples of common childhood disorders/problems that bring youngsters into a neuropsychologist's office include neurodevelopmental disorders such as ADHD, learning disabilities, intellectual disability, autism spectrum disorder, traumatic brain injury, toxic exposure (e.g., fetal alcohol spectrum disorder), cancer, chromosomal and genetic syndromes (e.g., neurofibromatosis, Fragile X, Klinefelter), cerebral palsy, and seizure disorders. Young to middle-aged adults are most often evaluated because of neurodevelopmental disorders, traumatic brain injuries, multiple sclerosis, cancer, infections (e.g., meningitis), substance use disorders, and psychiatric issues (e.g., mood, psychotic, and/or trauma-related symptoms). Older adults often find themselves in a neuropsychologist's office due to concerns about neurodegenerative conditions (e.g., Alzheimer's disease, Lewy body pathology,

---

[22] Some pediatric neuropsychologists see patients into their early 20s.

frontotemporal lobar degeneration), stroke/cerebrovascular disease, traumatic brain injuries, epilepsy, psychiatric symptoms, delirium (drastic fluctuations in awareness and attention), and hydrocephalus (a buildup of cerebrospinal fluid in the brain). Frequently, patients come in for an evaluation without a clear understanding of the underlying cause of their difficulties and part of our job is to help rule in or out certain problems.

Cognitive Domains Measured

Neuropsychologists are experts at assessing behavior. We use observation, interview questions, inventories, rating scales (e.g., self, parent, teacher), and cognitive tests to measure behavior, both subjectively and objectively. The subjective pieces (e.g., asking the person about changes in cognition or mood) are very useful.[23] However, what really sets us apart is our ability to *objectively* assess cognitive functioning. An individual's subjective sense of their abilities does not always line up with how they actually perform on testing (Jessen et al. 2014, 2020), which is why it is necessary to assess cognition at both levels. In other words, how do you know if your patient, who complains of forgetfulness, actually has a problem with memory? You test their memory! As such, much of what we collect are quantitative data, although qualitative information (e.g., flat affect, tangential speech, unusual behavior) is also highly valued.

   Below, we cover the specific areas, or cognitive domains, that modern neuropsychologists typically evaluate (see Table 1.1).[24] This is Neuropsychology 101, so we recommend reading this section closely. As you peruse each description, it is important to note that cognitive abilities are interrelated, and tests often tap into multiple domains (this is called "task impurity"). For example, in order to complete Part B of the Trail Making Test, a popular measure where the patient rapidly draws connecting lines between numbers and letters, they must demonstrate focused attention, visual scanning, motor sequencing, and rapid information processing, among other skills. A breakdown in any one of these abilities will lead to poor performance.

Six Core Cognitive Abilities

Attention We are constantly bombarded with a wealth of information from our senses, but our brains cannot process everything that we encounter at any given point in time. As conceptualized by neuropsychologists, attention is a cognitive

---

[23] For more information about qualitative aspects of neuropsychology, see the book, The Boston Process Approach to Neuropsychological Assessment, by Ashendorf and colleagues (2013).

[24] This is not meant to be an exhaustive list. The purpose is to give you a general framework about the primary domains assessed in neuropsychology. For a much more comprehensive review of the thousands of cognitive tests, read Sherman et al., 2020, Lezak et al. (2012), and Carlson and Geisinger (2017).

Table 1.1 A simple taxonomy of neuropsychological domains

Core	Supplemental
Attention	Performance validity
Processing speed	Academic achievement
Language	Sensorimotor
Visuospatial/visuoconstruction	Psychological functioning
Learning/memory	Intelligence
Executive functioning	

"spotlight" or "gatekeeper" that determines which information is processed and which information is ignored (for a fun demonstration of this phenomenon, Google *The Monkey Business Illusion*). With this in mind, attention is often broken down into several subcategories. The first is sustained attention – the ability to focus on a single stream of information for at least a moderate amount of time. When it goes awry, a person will be distractible any time a conversation, lecture, book, or other stimulus requires them to concentrate for more than a few minutes at a time. The second category is selective attention – the ability to *intentionally* block out irrelevant information and process relevant material. People with poor selective attention are pulled toward salient aspects of objects in their environment (e.g., a flashing billboard on the side of the highway) at the expense of more relevant stimuli (e.g., the cars around them traveling at 65+ MPH). Finally, there is divided attention – the ability to rapidly and efficiently shift attention between two different stimuli. Divided attention is frequently mislabeled as "multitasking." In reality, human beings are not capable of multitasking – that is, our brains cannot concurrently process two separate and complex streams of information. Instead, we rapidly switch back and forth from one stream to the other (Shallice et al. 2008). Highway driving is the perfect example; we do not simultaneously attend to the task of operating our vehicle *and* talking on the phone or sending text messages – we switch back and forth from one task to the other. That means that while we are attending to our phone, we are not attending to the cars around us, and we do so at our own (and others') peril.

Processing Speed Now that we have established what it means to attend to stimuli, we will consider the rate at which we do so. People vary greatly in their information processing speed – the rapidity with which they analyze and respond to environmental stimuli to which they have attended. Processing speed is tightly correlated with age such that the older we are (after our mid-twenties or so), the slower we process information (Eckert 2011). This is a big part of the reason why elite Formula One drivers and National Football League quarterbacks tend to be in their 20s and 30s.

Language We all have a sense as to what language is – on a basic level, it is a system of communicating with other members of the same species. However, when neuropsychologists consider language, we organize it into components that are tied to brain networks – receptive language (the ability to understand), expressive language (the ability to communicate our thoughts to others), and "naming" (word retrieval and expression) are a few of the most commonly assessed skills. Although attention is the most basic cognitive ability (everything else depends on it), language is not far behind. It is very difficult to accurately assess other areas of cognitive functioning such as mood, memory, and problem solving when the person across from you cannot comprehend you and/or express themselves to you.

Visuospatial/Visuoconstructional Abilities These skills are a bit less intuitive than the last three. What we are describing here are cognitive abilities that allow us to understand spatial relationships. Because we do not localize sounds well and our senses of smell and touch are relatively rudimentary (compared to other animals), our spatial skills are primarily visual (hence, "*visuo*spatial"). So, what exactly do we mean by spatial relationships? Think about tasks performed by engineers or architects: envisioning the position of an object in three-dimensional space, rotating it, creating a mental map of its interior, and then manipulating that map. These are the core aspects of visuospatial skills. Then, when we say "visuoconstructional skills," we mean visuospatial + a motor task (e.g., building a Millennium Falcon™ out of LEGOs®). When these skills go awry in a typical person, we observe that they have trouble navigating their environment. They are at risk of becoming lost and confused when driving to the store, when walking around their neighborhood, or even when ambulating around their home. Interestingly, in the world of modern technology, visuospatial skills are far less essential than they were throughout human history. We have GPS built into our cars and cell phones, so we do not always need to use our own mental maps to navigate the world as our predecessors, the hunter-gatherers, did. If you are visuospatially challenged as I (Ryan) am, you will be very thankful for this fact.[25]

Learning/Memory After we attend to information, our brains process and *encode* it into short-term memory storage. From there, it is *consolidated,* or transformed, from short-term to long-term memory. Later (days, weeks, months, years), it is *retrieved* when our brains activate the particular circuit that corresponds to that memory. For example, at some point in your early education, you probably learned many facts that you may or may not be using in your daily life (e.g., $12 \times 7 = 84$; the second President of the United States was John Adams). Each time you learned one of these tidbits, the structure of your brain was being slightly altered – neurons were expanding and branching out ("dendritic arborization") and releasing neurotransmitters at synapses, thereby forming new connections with other neurons and allowing your brain to accommodate the new information. Now, if I were to ask you, "How many quarts are in one gallon?" and you were to answer, "four," this happened because executive control processes in your brain searched your memory store, rapidly located the corresponding memory circuit, and pulled it into your

---

[25] If you are interested in learning more about visuospatial skills, check out the book, *Mind in Motion* by Barbara Tversky.

conscious awareness. Of course, this complex set of memory processes can go awry at each and every stage. You may not encode information that you attended to, you may initially encode it but fail to consolidate it, and you may have it fully stored in long-term memory but be unable to retrieve it in a particular circumstance. Neuropsychologists test and parse out memory failures at each of these stages because memory is integral for everyday functioning and because different disease processes/injuries impact memory in unique ways.

Executive Functioning We started with the most basic and foundational cognitive construct (attention) and we have been working our way up the hierarchy of complexity. We have now reached the zenith – the highest level of thinking – which we call "executive functioning." This is a heterogeneous *family* of abilities. We often explain it to patients as "the CEO of your mind" – it is the boss who solves problems, makes decisions, shifts attention between multiple projects, shows restraint where necessary, and displays mental flexibility and dexterity in making sense of the world. Executive functions rely on the functioning of all of the lower-level systems discussed earlier. They are also very susceptible to becoming impaired when someone sustains a brain injury, leading to fascinating (and tragic) cases of "dysexecutive" behavior such as frank disinhibition, where a person loses all sense of self-awareness (their "filter") and becomes prone to cursing, making sexually inappropriate advances, urinating in public, etc.

Supplemental Domains

Performance Validity This is not a "cognitive domain," per se, but it is an essential consideration in each and every neuropsychological evaluation. What we mean by "performance validity" is the extent to which a patient's cognitive scores represent their true abilities. This is imperative because we cannot make accurate judgments about a person's cognitive strengths and weaknesses if their low scores (interpreted as weaknesses) result from something other than true deficits. There are myriad reasons why this may be the case: a person may be severely underslept, they may be acutely confused, they may be highly medicated, they may be in acute pain and/or highly distressed, or they may be intentionally performing poorly in order to appear impaired and receive some type of secondary gain (e.g., workers' compensation benefits). Because of the importance of this issue, neuropsychologists have a plethora of techniques to detect performance *invalidity*, including behavioral observations and specialized tests.

Academic Achievement Here we are referring to scholastic performance, especially with regard to core abilities such as reading, writing, and math. We think of these skills as cognitive "aftermarket add-ons," (think of a standard Volkswagen Jetta with tinted windows and a turbocharged engine) because our brains did not evolve the abilities to perform algebra or write out sentences onto paper. Instead, these skills have become important to our functioning due to cultural and industrial advances. In the world of neuropsychological evaluations, academic achievement is particularly important in the assessment of learning disorders. Some children have great difficulty developing and fine-tuning the building blocks necessary for successful performance in school (e.g., phonological awareness, orthographic processing), likely because their brains are less adept at adding the tinted windows and turbocharged engine of reading and writing. Fortunately, there are tests that can detect early problems in academic achievement and, once a child's particular set of strengths and weaknesses is identified, the neuropsychologist can then recommend personalized interventions to bring the child closer to grade level in a particular skill. This is incredibly important because the slow development of basic academic skills can constrain a person throughout the rest of their educational and occupational lives.[26]

Sensorimotor Like everything we do, think, and experience, sensorimotor skills are controlled by the brain. However, they are typically conceptualized as distinct from cognitive abilities and are generally the wheelhouse of other experts such as ophthalmologists, audiologists, physical therapists, kinesiologists, and neurologists. Still, because sensory impairments can masquerade as cognitive deficits (e.g., hearing loss can be mistaken for memory loss) and because motor impairments are present in a number of diseases that impact the brain and cognition (e.g., Parkinson's disease, Tourette's disorder), it is often important for neuropsychologists to briefly assess basic sensation and movement. We do this by asking patients to complete tests of basic visual acuity (e.g., "read the letters on this page from top to bottom"), olfaction (e.g., "what does this scratch and sniff smell like?"), rapid movements (e.g., "tap your index finger as quickly as you can for 10 seconds"), fine-motor control (e.g., "rotate and place each of these pegs into the small holes as quickly as you can"), and strength (e.g., "squeeze this device as hard as you can"). With regard to sensation, deficits such as hearing loss and visual field deficits can be identified and accommodated for in the neuropsychological evaluation so that cognition, rather than vision or hearing, is measured. Neuropsychologists might also provide recommendations for patients to receive specialist assessments for vision or hearing loss if it appears that they have untreated problems in these areas. In terms of motoric abilities, bilateral or asymmetric weaknesses can provide information about possible brain pathology and can help us determine what someone can and cannot do (e.g., go back to work, drive a car), as well as what types of recommendations might assist them in their daily lives.

Psychological Functioning Most people think of social/emotional functioning as completely distinct from cognitive functioning. The former refers to someone's self-confidence, their mood, their propensity to seek out other people, their compassion,

---

[26] If you want to learn more about learning disorders, we recommend the book, *Diagnosing learning disorders: From science to practice, 3rd ed.,* (2019) by Pennington, McGrath, & Peterson. Also, see www.NavNeuro.com/61.

their level of shyness, etc., while the latter refers to their thinking skills. In reality, emotions and cognition are very closely connected in the brain. When someone is depressed, for example, their attention and memory systems function differently and less effectively. Another common psychological state with a major impact on cognition is anxiety. Think back to a time when you experienced performance anxiety – maybe it was immediately before you delivered a presentation in front of your classmates or before you performed in a play or musical. In the grips of crippling anxiety, were you at your very best? Do people dance, sing, take tests, play sports, or give speeches at the highest level when they are paralyzed by anxiety? Of course not. Consequently, when neuropsychologists are evaluating cognition, we also measure psychological functioning. It is not uncommon for someone to think that their memory is being stolen away by an untreatable degenerative condition such as Alzheimer's disease when, in reality, their attention and memory have been constrained by depression. The latter is treatable, and with the alleviation of psychological symptoms, cognitive skills can improve greatly. Therefore, neuropsychologists play an important role in the identification and treatment of psychological distress in people with subjective cognitive decline.

Intelligence There is a long, rich history of research on intellectual abilities in humans. However, we will not go into detail on this topic here for several reasons. First, "intellectual abilities" overlap greatly with "cognitive abilities," and everything that can be objectively measured and labeled as "intelligence" can also be called "cognition." As you may have noticed, we prefer "cognition" because we believe that it is less biased and more useful. Second, and relatedly, the term "intelligence" is often misused and misunderstood. Most people think of it as a synonym for "smart" and use it very loosely; we think that this usage is unhelpful. When we formally measure intelligence, we are only capturing a narrow range of abilities (sometimes called Mental Abilities as Measured by Intelligence Tests, or MAMBIT; Stanovich 2009) and we are missing important real-world skills such as emotional intelligence, musical intelligence, existential intelligence, and bodily-kinesthetic intelligence (Gardner 1983, 1993). Third, "intelligence" has sometimes been used to demean, judge, and ostracize whole groups of people in tragic ways (e.g., the eugenics movement), while "cognition" has not been guilty of such travesties. With all of this said, the past 100+ years of research on intelligence have produced helpful insights into human thinking and we think that this research evidence has utility (Flanagan and Harrison 2018). We simply prefer to include this research under the umbrella of cognitive abilities rather than that of intellectual abilities.

How Is Neuropsychology Different from Related Fields?

Contemporary neuropsychology can easily be confused with related clinical fields such as neurology and psychiatry. There are many similarities between neuropsychology and these other professions, and we generally work with the same patients in a collaborative fashion in order to arrive at a diagnosis and manage or treat the problem. Furthermore, we all have similar roots, splitting off from the parent fields (namely, neuroscience and philosophy) at different junctures. As noted above, one helpful framework is to think of the distinction as being due to different areas of knowledge, patient groups, and clinical tools. We will describe some of the main similarities and differences between neuropsychology and related fields here (also, see Table 1.2).

Neurology

Neurologists are medical doctors who have extensive training in biology and physiology, with a particular focus on the central nervous system – the brain and spinal cord. They treat many of the same patients as neuropsychologists – patients with

Table 1.2 Comparing and contrasting allied health professions to neuropsychology

Discipline	Similarities	Differences
Neurology	Serves patients with neurological conditions	Attended medical school rather than graduate school
		Prescribes medications and medical procedures
		Conducts neurological examinations
		Less training in mental illness, cognitive and psychological assessment, statistics, and research methods
Psychiatry	Serves patients with psychiatric conditions	Attended medical school rather than graduate school
		Prescribes medications and medical procedures
		Less training in cognitive and psychological assessment, statistics, and research methods
Occupational therapy	Serves patients with cognitive and emotional symptoms	Assesses and treats functional (rather than neuropsychiatric[a]) issues
		Primarily master's-level training
Speech-language pathology	Serves patients with language disorders	Primarily focused on language abilities
		Conducts extensive interventions for speech and communication
		Primarily master's-level training
Psychometry	Serves patients with neuropsychiatric conditions	Does not interpret tests, make diagnoses, or provide feedback to patients
	Administers and scores neuropsychological tests	Works under a psychologist's license
		Primarily bachelor's or master's-level training

[a]The term "neuropsychiatric" helps resolve the false neurology-psychiatry dichotomy by creating a larger umbrella within which traditional "neurological" (e.g., brain tumor) and traditional "psychiatric" (e.g., anxiety) disorders exist

brain diseases such as Alzheimer's disease, stroke, multiple sclerosis, movement disorders, and epilepsy – but they use different tools. They focus on neuroimaging such as CT scans and MRIs, as well as electroencephalography (EEG), blood tests, and lumbar punctures (spinal taps). They also test reflexes and cranial nerve functioning and they write prescriptions for medications. By contrast, neuropsychologists are psychologists by training, not physicians. That means that we attend graduate school rather than medical school and we specialize in human thinking, emotion, and behavior. With few exceptions, we do not prescribe medications and we do not typically order or directly interpret brain scans and other medical procedures. Instead, we develop extensive knowledge and skills in mental illnesses (psychopathology) as well as in the assessment of cognition and personality traits, and we receive training in psychotherapy and cognitive interventions. Finally, in contrast to medical school, we receive training in statistics and research methods during graduate school, and this greatly enhances our abilities as researchers. In a nutshell, compared to neurologists, neuropsychologists assess and treat problems with personality, mood, and cognitive skills, whereas neurologists handle physical examinations, neuroimaging, and medications.

As a neurologist, I routinely benefit from the ability to access the knowledge and skills of our neuropsychology team. A particularly memorable patient had recovered from a traumatic brain injury but was struggling to reintegrate to his work environment. His consultation with a neuropsychologist revealed difficulty with processing parallel input streams and led to concrete behavioral adaptations that radically improved his performance. I was so happy that they were there for him and for me!

– Sean Evans, MD

Early diagnosis of neurodegenerative diseases represents a complex clinical puzzle for cognitive neurologists to solve. Trying to solve these clinical puzzles requires exceptional tools. An adept neuropsychologist, with in-depth testing and insightful interpretation, can be a key piece to helping us understand which large scale brain network is likely involved. This knowledge helps improve early diagnosis and understand how these diseases evolve.

– Ryan Townley, MD

Psychiatry

Similar to neurologists, psychiatrists are also medical doctors, but they specialize in treating psychopathology rather than traditional brain-related problems. This is actually a false dichotomy because mental illnesses occur in the brain just as a tumor or a stroke occurs in the brain; however, this is how medicine has been conceptualized for years and this terminology is still in use today.[27]

Psychiatrists are experts in the use of psychotropic medications such as antidepressants, antipsychotics, mood stabilizers, and anti-anxiety medications; these are

---

[27] In fact, neurology and psychiatry are combined into one board in the US, The American Board of Psychiatry and Neurology. Also, the American Neuropsychiatric Association reflects influence from both of these fields: https://www.anpaonline.org/

all used specifically to treat the symptoms of mental illnesses. Psychiatrists understand important aspects of these medications such as correct dosage, potential side effects, and complex interactions between different agents. They spend a significant amount of time identifying the optimal medication regimen for each individual patient, and this is a complicated task because everyone's biology is unique. Psychiatrists also occasionally perform psychotherapy and other interventions, but their bread and butter is a brief medication consultation with their patients. So, while psychiatrists and neuropsychologists often work with the same patients – people with depression, anxiety, bipolar disorder, schizophrenia, and other mental illnesses – the psychiatrist's primary tools are medications, while the neuropsychologist's primary tools are cognitive tests.

Occupational Therapy

Contrary to what is suggested by their title, occupational therapists (OTs) are not strictly focused on helping people go back to work. They do address vocational issues, but they do it within the broader context of assessing and improving a patient's functional abilities – i.e., how well the patient performs the tasks that are needed to take care of themselves and fulfill their daily obligations. As an example, an OT might assess whether or not a patient can prepare a meal, write a check, drive their car, or manage their medications; next, the OT generates a plan to strengthen these abilities if necessary. This is incredibly important because a patient's functional problems can be a safety hazard, potentially leading to a car accident, malnutrition, a fall, or a house fire. Improving activities of daily living also helps people remain independent and in their homes for longer and enhances their overall quality of life.

Neuropsychologists and OTs often work together in caring for a patient. The neuropsychologist measures the patient's cognitive abilities, while the OT measures and works to improve the patient's functional skills. For many people with neurological illnesses, the changes in their brain typically first impact their thinking skills and later cause a decline in their ability to carry out everyday activities.

Speech-Language Pathology

Speech-language pathologists (SLPs) work with patients who have deficits in language and communication, one of the cognitive domains described above. For example, if a stroke causes significant language impairment (aphasia), an SLP can assess the extent and type of impairment and then treat the problem. They also commonly work with children with neurodevelopmental disorders and with individuals with difficulty swallowing. Neuropsychologists and SLPs often work together in rehabilitation hospitals; in these settings, the neuropsychologist typically assesses the nature and extent of the language problem, as well as the impact on other areas of cognitive functioning. The SLP then uses the findings from the neuropsychological evaluation to create a personalized treatment regimen where they help the patient improve the language deficits.

Psychometry

Psychometrists are trained in administering and scoring various cognitive and psychological tests. Many neuropsychologists work with a psychometrist – the neuropsychologist performs the medical records review and gathers background information from the patient, and then the psychometrist administers and scores the tests. Later, the neuropsychologist writes the report, consults with allied health professionals as necessary, and provides the patient with feedback about the results. Because they are not involved in diagnosis or interpretation, psychometrists typically have a bachelor's or master's degree (as opposed to a doctorate), and they can pursue certification in psychometry as well.[28] Psychometrists are very skilled in terms of developing rapport with patients, observing and recording behavior, handling difficult clinical situations (e.g., when a patient has visual impairments, narcolepsy, or psychotic symptoms), and administering and scoring dozens of complex neuropsychological tests. Neuropsychologists have great respect and appreciation for their psychometrist colleagues, and psychometrists help improve the efficiency of neuropsychologists. If you want to learn more, visit the National Association of Psychometrists' website, https://www.napnet.org/.

Conclusion

We have come a long way in understanding the causes of behavior, as well as how to identify and manage maladaptive neurocognitive abnormalities. It took thousands of years before the brain was widely recognized as the seat of thoughts, emotions, and actions, and several hundred more years before specific brain networks were linked to specific behavioral outputs in a systematic and objective manner. A major goal for this chapter has been to describe how the discipline of neuropsychology emerged from this cacophony of intellectual discoveries and insights.

And yet there is much still to be learned. For example, we have only recently begun to realize that the localization model is only partially true – that is, that the connections among brain regions and parallel processes are at least as important as the specialized regions of interest (e.g., your vision would be impaired if the connections

---

[28] Requirements vary by state.

to and from your occipital lobe were disrupted).[29] Over time, new advancements in neuromonitoring will allow us greater temporal and spatial resolution as we examine the structure and function of the human brain. We will better understand the genetic underpinnings of diseases and developmental issues. Precision medicine will help us tailor interventions to each individual. Technological advances will allow clinicians to share data, to optimize the efficiency of the modern neuropsychological evaluation, and to study patients outside of the clinic – for example, by receiving data from wearables or other "smart" devices. Indeed, some neuropsychologists believe that we are in the midst of a revolution in the field, something that Dr. Robert Bilder terms "Neuropsychology 3.0" (Bilder 2011).[30,31]

Given the relative youth of the field and the rapidity of technological and scientific progress, the answer to the question "What is neuropsychology?" will likely be a bit different by the time you become a neuropsychologist. We currently stand on the shoulders of giants, but many more such titans of science and clinical practice are needed to accomplish the tasks that will lead neuropsychology into the future. We're counting on you!

References

Aciduman, A., Arda, B., Özaktürk, F. G., & Telatar, Ü. F. (2009). What does Al-Qanun Fi Al-Tibb (The Canon of Medicine) say on head injuries? *Neurosurgical Review, 32,* 255–263.

Al-Rodhan, N. R., & Fox, J. L. (1986). Al-Zahrawi and Arabian neurosurgery, 936–1013 AD. *Surgical Neurology, 26*(1), 92–95.

Aristotle. (1912). *De Partibus Animalium* (W. Ogle, Trans., & J. A. Smith & W. D. Ross, Ed.). Oxford. (Original work published ca. 350 BCE).

Armstrong, K. E., Beebe, D. W., Hilsabeck, R. C., & Kirkwood, M. W. (2019). *Board certification in clinical neuropsychology: A guide to becoming ABPP/ABCN certified without sacrificing your sanity* (2nd ed.). Oxford University Press.

Barth, J. T., Pliskin, N., Axelrod, B., Faust, D., Fisher, J., Harley, J. P., et al. (2003). Introduction to the NAN 2001 definition of a clinical neuropsychologist. *Archives of Clinical Neuropsychology, 18*(5), 551–555.

Benton, A. (2000). *Exploring the history of neuropsychology: Selected papers.* Oxford University Press.

Bilder, R. M. (2011). Neuropsychology 3.0: Evidence-based science and practice. *Journal of the International Neuropsychological Society, 17*(1), 7–13.

---

[29] The full web of connections between all brain cells is called the "connectome." The Human Connectome Project launched in 2009 as a large-scale attempt to comprehensively map all of the brain's connections. As of the time of writing this book, the project is still on-going. We recommend Sebastian Seung's book, *Connectome: How the brain's wiring makes us who we are* (2012) on this topic.

[30] See Bilder and Reise (2019) for a more in-depth discussion of technology that could revolutionize the field of neuropsychology. See NavNeuro episodes 33, 35, and 37 for discussions with Dr. Bilder.

[31] If you are looking for a book on innovation in our field, check out, *The role of technology in clinical neuropsychology* (2019) by Kane & Parsons.

Bilder, R. M., & Reise, S. P. (2019). Neuropsychological tests of the future: How do we get there from here? *The Clinical Neuropsychologist, 33*(2), 220–245.

Board of Directors. (2007). American Academy of Clinical Neuropsychology (AACN) practice guidelines for neuropsychological assessment and consultation. *The Clinical Neuropsychologist, 21*(2), 209–231.

Bressler, S. L., & Menon, V. (2010). Large-scale brain networks in cognition: Emerging methods and principles. *Trends in Cognitive Sciences, 14*(6), 277–290.

Broca, P. (1865). Sur le siège de la faculté du langage articulé (15 juin). *Bulletins de la Société Anthropologque de Paris, 6*, 377–393.

Bruce, D. (1985). On the origin of the term "neuropsychology". *Neuropsychologia, 23*(6), 813–814.

Carlson, J. F., & Geisinger, K. F. (Eds.). (2017). *The twentieth mental measurements yearbook*. Buros Center for Testing.

Drouin, E., & Péréon, Y. (2019). Dax versus Broca. *The Lancet Neurology, 18*(10), 920.

Eckert, M. A. (2011). Slowing down: Age-related neurobiological predictors of processing speed. *Frontiers in Neuroscience, 5*, 25.

Faria, M. A. (2015). Neolithic trepanation decoded-A unifying hypothesis: Has the mystery as to why primitive surgeons performed cranial surgery been solved? *Surgical Neurology International, 6*, 72.

Finger, S. (1994). History of neuropsychology. In D. W. Zaidel (Ed.), *Neuropsychology: Handbook of perception and cognition* (2nd ed., pp. 1–28). Academic.

Flanagan, D. P., & Harrison, P. L. (2018). *Contemporary intellectual assessment: Theories, tests, and issues* (4th ed.). The Guilford Press.

Gardner, H. (1983). *The theory of multiple intelligences*. Basic Books.

Gardner, H. (1993). *Multiple intelligences: The theory in practice*. Basic Books.

Geschwind, N. (1975). The borderland of neurology and psychiatry: Some common misconceptions. In D. F. Benson & D. Blumer (Eds.), *Psychiatric aspects of neurologic disease* (Vol. 1). Grune & Stratton.

Golden, C. J., Purisch, A. D., & Hammeke, T. A. (1979). *Luria-Nebraska neuropsychological test battery: A manual for clinical and experimental uses*. University of Nebraska Press.

Greenaway, M. C., Duncan, N. L., & Smith, G. E. (2013). The memory support system for mild cognitive impairment: Randomized trial of a cognitive rehabilitation intervention. *International Journal of Geriatric Psychiatry, 28*(4), 402–409.

Grote, C. L., Butts, A. M., & Bodin, D. (2016). Education, training and practice of clinical neuropsychologists in the United States of America. *The Clinical Neuropsychologist, 30*(8), 1356–1370.

Hannay, H. J., Bieliauskas, L. A., Crosson, B. A., Hammeke, T. A., Hamsher, K. deS., & Koffler, S. P. (1998). Proceedings of the Houston conference on specialty education and training 802 in clinical *neuropsychology*. *Archives of Clinical Neuropsychology, 13*(2), 157–158.

Harlow, J. M. (1848). Passage of an iron bar through the head. *The Boston Medical and Surgical Journal, 39*(20), 389–393.

Harlow, J. M. (1868). *Passage of an iron bar through the head*. Publications of the Massachusetts Medical Society.

Huckans, M., Hutson, L., Twamley, E., Jak, A., Kaye, J., & Storzbach, D. (2013). Efficacy of cognitive rehabilitation therapies for mild cognitive impairment (MCI) in older adults: Working toward a theoretical model and evidence-based interventions. *Neuropsychology Review, 23*(1), 63–80.

INS-Division 40 Task Force. (1987). Reports of the INS-Division 40 task force on education, accreditation, and credentialing. *The Clinical Neuropsychologist, 1*, 29–34.

Jessen, F., Amariglio, R. E., Van Boxtel, M., Breteler, M., Ceccaldi, M., Chételat, G., et al. (2014). A conceptual framework for research on subjective cognitive decline in preclinical Alzheimer's disease. *Alzheimer's & Dementia, 10*(6), 844–852.

Jessen, F., Amariglio, R. E., Buckley, R. F., van der Flier, W. M., Han, Y., Molinuevo, J. L., et al. (2020). The characterisation of subjective cognitive decline. *The Lancet Neurology, 19*(3), 271–278.

Kamp, M., Tahsim-Oglou, Y., Steiger, H.-J., & Hänggi, D. (2012). Traumatic brain injuries in the ancient Egypt: Insights from the Edwin Smith Papyrus. *Journal of Neurological Surgery Part A: Central European Neurosurgery, 73*(4), 230–237.

Kane, R. L., & Parsons, T. D. (Eds.). (2019). *The role of technology in clinical neuropsychology.* Oxford University Press.

Kolb, B., & Whishaw, I. Q. (2015). *Fundamentals of human neuropsychology* (7th ed.). Worth Publishers.

Konstantine, P. P., & Peter, K. P. (2015). The ancient Greek discovery of the nervous system: Alcmaeon, Praxagoras and Herophilus. *Journal of Clinical Neuroscience, 29*, 21–24.

Kostyanaya, M. I., & Rossouw, P. (2013). Alexander Luria – Life, research and contribution to neuroscience. *International Journal of Neuropsychotherapy, 1*(2), 47–55.

Lezak, M. D., Howieson, D. B., Bigler, E. D., & Tranel, D. (2012). *Neuropsychological assessment* (5th ed.). Oxford University Press.

Luria, A. R. (1966). *Higher cortical functions in man.* Basic Books, Inc.

Macmillan, M. (2000). *An odd kind of fame: Stories of Phineas Gage.* The MIT Press.

Milner, B., Corkin, S., & Teuber, H. (1968). Further analysis of the hippocampal amnesic syndrome: 14-year follow-up study of HM. *Neuropsychologia, 6*(3), 215–234.

Mirzaa, G. M., & Poduri, A. (2014). Megalencephaly and hemimegalencephaly: Breakthroughs in molecular etiology. *American Journal of Medical Genetics Part C: Seminars in Medical Genetics, 166*(2), 156–172.

Nanda, A., Khan, I. S., & Apuzzo, M. L. (2016). Renaissance neurosurgery: Italy's iconic contributions. *World Neurosurgery, 87*, 647–655.

Pennington, B. F., McGrath, L. M., & Peterson, R. L. (2019). *Diagnosing learning disorders: From science to practice.* The Guilford Press.

Rabin, L. A., Paolillo, E., & Barr, W. B. (2016). Stability in test-usage practices of clinical neuropsychologists in the United States and Canada over a 10-year period: A follow-up survey of INS and NAN members. *Archives of Clinical Neuropsychology, 31*(3), 206–230.

Sacks, O. (1985). *The man who mistook his wife for a hat and other clinical tails.* Summit Books.

Sacks, O. (1995). *An anthropologist on mars: Seven paradoxical tales.* Vintage Books.

Shallice, T., Stuss, D. T., Picton, T. W., Alexander, M. P., & Gillingham, S. (2008). Mapping task switching in frontal cortex through neuropsychological group studies. *Frontiers in Neuroscience, 2*(1), 79–85.

Shephard, B. (2015). Psychology and the Great War, 1914–1918. *The Psychologist, 28*(11), 944–946.

Sherman, E. M. S., Tan, J. E. E., & Hrabok, M. (2020). *A compendium of neuropsychological tests: Fundamentals of neuropsychological assessment and test reviews for clinical practice* (4th ed.). Oxford University Press.

Stanovich, K. E. (2009). *What intelligence tests miss: The psychology of rational thought.* Yale University Press.

Tversky, B. (2019). *Mind in motion: How action shapes thought.* Basic Books.

Twamley, E. W., Vella, L., Burton, C. Z., Heaton, R. K., & Jeste, D. V. (2012). Compensatory cognitive training for psychosis: Effects in a randomized controlled trial. *The Journal of Clinical Psychiatry, 73*(9), 1212–1219.

Twamley, E. W., Jak, A. J., Delis, D. C., Bondi, M. W., & Lohr, J. B. (2014). Cognitive Symptom Management and Rehabilitation Therapy (CogSMART) for veterans with traumatic brain injury: Pilot randomized controlled trial. *Journal of Rehabilitation Research & Development, 51*(1), 59–70.

Wernicke, C. (1874). *Der aphasische Symptomencomplex: Eine psychologische Studie auf anatomischer Basis.* Cohn & Weigert.

Chapter 2

Why Neuropsychology?

> My tail is wagging so hard every day I go to work. I love it. I think we are in the best field ever. The more we learn about every single physical system in our body, the more we understand that it all affects our cognition. And the more we learn about neuroscience, the more we understand the connection between our emotions and our thought processes... we are in an amazing field.
>
> – Karen Postal, PhD, ABPP-CN

As we mentioned in the Preface, deciding which career path to take requires thoughtful consideration of all of the reasons why you might pursue different professions. One option for going about the decision-making process is to create a spreadsheet (riveting stuff!). Each row can represent a different occupation (e.g., "licensed clinical social worker," "psychiatrist," "rehabilitation psychologist," "neuropsychologist," "neurologist"), and each column can be filled in with various dimensions of work that you consider relevant and important in your life (e.g., "potential for creative problem solving," "the ability to help people," "upward mobility"). We provide a cursory example in Table 2.1 below. In this chapter, we will help you fill in the cells in the "neuropsychology" section.

Reasons for choosing neuropsychology vary greatly and we believe that it is precisely this diversity that makes the profession so appealing to people of different cultural, academic, and social backgrounds. We spoke to a number of our friends and colleagues about this topic and noticed the following trends: *having a positive impact, inspiration from loved ones with brain damage, developing brain-behavior expertise, a melting pot of knowledge and skills, novelty and intellectual engagement, prestige, financial compensation,* and *job security.* In this chapter, we do a deep dive into each of these themes in order to give you a taste as to why many current neuropsychologists and trainees selected this field. We readily admit that we are biased here, but we feel very strongly that neuropsychology has an immense amount

© Springer Nature Switzerland AG 2021
J. A. Bellone, R. Van Patten, *Becoming a Neuropsychologist*,
https://doi.org/10.1007/978-3-030-63174-1_2

Table 2.1 Spreadsheet method for career decision-making

Occupation	Have a positive impact	Adequate compensation	Spend significant time with patients
Social work			
Psychiatry			
Neuropsychology			

to offer someone who is motivated and interested in the brain and behavior. Our personal experiences, as well as those from our friends and colleagues, suggest that people who choose neuropsychology as a career path end up thoroughly enjoying their work.

As you read about all possible reasons to pursue this field, we urge you to begin contemplating your own personal, "Why?" It is crucial that you think deeply and carefully about this question. Developing and cultivating a strong set of your own personal motives and values will benefit you greatly in the years to come. This confidence and strength can serve as an endless source of inspiration that will sustain you through the years of challenging undergraduate and graduate work. In our experience, even difficult quizzes and tests, busy work schedules, and complex clinical/scientific work can be experienced as fun and engaging if you possess the inner conviction that you are on the right path, moving toward a fulfilling and meaningful career.

Without further ado, here are some of the most commonly cited inclinations, motivations, and inspirations that lead people down the path to becoming brain-behavior experts:

Having a Positive Impact

At the very top of many neuropsychologists' list of reasons for their career choice is, quite simply, "helping people." There are countless careers that involve making a difference in people's lives (e.g., social worker, physician, massage therapist, teacher, school counselor, politician, veterinarian), and all of the helping professions are important. However, there are specific aspects of neuropsychology that allow us to benefit people in unique ways and to make a difference in the lives of those who are especially disadvantaged and vulnerable.

> Throughout high school, I worked in a nursing home for patients with dementia. Witnessing cognitive and functional decline in these patients sparked a desire to help and a curiosity for comprehending factors contributing to the development of dementia.
>
> – Liselotte de Wit, MS

> We are, as neuropsychologists, optimally positioned for having an impact on healthcare in the next century, since many of the diseases that are going to be faced will have a cognitive component.
>
> – Michael Parsons, PhD, ABPP-CN

As neuropsychologists, our job allows us to have a direct impact on people's lives, often through clinical care, research, or teaching. When something goes awry with people's typical cognitive abilities, such as when a stroke impacts a person's language and communication, or when Alzheimer's disease affects someone's memories of their loved ones, people turn to neuropsychologists for answers and guidance. There is something profoundly special about this responsibility. As neuropsychologists, we are entrusted with the duty to measure, explain, and improve people's thinking – to take the inner workings of our patients' minds and bring them to light, so that they and their loved ones can better understand their strengths and limitations.

One aspect of our job as clinical neuropsychologists that many other helping professions are not privy to is our ability to spend an extended period of time with each patient (often 3–6 hours for a single outpatient session). Such a significant chunk of time dedicated to a single person provides us with the opportunity to learn our patients' personal stories and to build rapport. It allows us to stitch together a detailed background and symptom history, gather information about their personality profile and thinking styles, and comprehensively elucidate their cognitive abilities.

We (John and Ryan) spend a great deal of our time working with older adults and we hear, over and over again, that many people's greatest fears are memory loss, cognitive decline, and dementia. We strongly empathize with these fears and we recognize that preserving/maintaining cognitive functions is a top priority for many people as they age. Due to advances in medicine such as treatments for cancer and HIV/AIDS, people are living longer. However, what is even more important than simply living longer is preserving physical and mental health as we age. In particular, the maintenance of one's cognitive abilities is vital to excellent *quality* of life, including happiness, wellness, and flourishing year after year, decade after decade (Medalia and Erlich 2017). Our job as neuropsychologists puts us on the front lines with respect to investigating, assessing, and treating many common brain diseases that threaten to strip people's cognitive abilities away from them.

On the other end of the age spectrum, there are many neuropsychologists who work with children and who find this role to be incredibly rewarding. Pediatric neuropsychologists frequently talk about how much they value the fact that, on a daily basis, they are able to make a direct impact in the life of a child and their family. As we mentioned in Ch. 1, pediatric neuropsychologists often measure and treat cognition in children who have learning disabilities, ADHD, autism spectrum disorder, intellectual disability, traumatic brain injuries, epilepsy, cancer, and other brain insults. Importantly, not only are these clinicians affecting the child in the present but they are also likely making a difference in that child's developmental trajectory and future adult life. Many adults walking around and contributing to society have their pediatric healthcare providers to thank for the excellent treatment and care they received for brain diseases/injuries they endured as children.

I appreciated neuropsychology's ability to develop a very unique picture of each patient, and to generate information and recommendations that can have a powerful impact on patients.

– Megan Spencer, PhD

To Dr. Spencer's point, there is published research on the positive impact that neuropsychologists have on the people we serve. For example, the Neuropsychology Outcome Satisfaction Initiative (NOSI) was a multisite outcome survey of a group of patients and their caregivers who completed neuropsychological evaluations. In other words, people who had previously been seen by clinical neuropsychologists answered survey questions about the benefits of these experiences. As a result of the evaluation, patients reported a significant improvement in their understanding of how their cognitive symptoms related to the diagnosis, their understanding of what to expect in the upcoming months, and their ability to cope with their current symptoms. Caregivers reported significant improvements in their ability to cope with the patients' symptoms and they indicated that they left the appointment with a good treatment plan for managing those symptoms. Both patients and caregivers agreed that the evaluation was useful for developing strategies to accomplish tasks and for working around the patients' cognitive difficulties. They also agreed that the neuropsychological assessment process helped them identify ways to reduce their stress level and that the evaluation and resultant feedback helped them make better long-term plans. Overall, the survey provides strong evidence that people do benefit from working with neuropsychologists.[1] In addition to patients and caregivers, the physicians who refer patients to neuropsychologists have also reported high levels of satisfaction with, and usefulness of, our evaluations (Temple et al. 2006).

****

A fascination with the study of human behavior, particularly abnormal behavior, led me to pursue Psychology as a major. In my final year, I took a research methods course and discovered my passion for research.

– April Thames, PhD

I engage in research to identify important breakthroughs in understanding brain-behavior relationships and in developing effective interventions.

– Geoffrey Tremont, PhD, ABPP-CN

We have discussed the clinical utility of neuropsychology, but it is also important to note that a large proportion of neuropsychologists devote their time to scientific work. In our role as researchers, we can publish scientific papers that have a positive impact on hundreds, thousands, or millions of people. For example, a neuropsychologist at the University of Florida, Dr. Glenn Smith, developed the Healthy Action to Benefit Independence and Thinking (HABIT) program, which is a 10-day, 50-hour intervention that has shown great promise in improving mood and overall

---

[1] There are many similar studies showing the benefits of neuropsychology. For example, see Allott et al. (2011); Hilsabeck et al. (2014); Mahoney et al. (2017); Rosado et al. (2017); Stark et al. (2014); VanKirk et al. (2013); Watt and Crowe (2018).

functioning in older adults with mild cognitive impairment (Greenaway et al. 2008). Neuropsychologists have also reported on scientific investigations showing that our techniques can improve cognitive skills, mood, functioning, and quality of life in people with schizophrenia (Twamley et al. 2012, 2017; http://www.cogsmart.com), can treat depression and anxiety in people with hoarding disorder (Ayers et al. 2018), and can reduce long-term symptoms in people who sustained concussions (McCrae 2008). These are just a few examples of the plethora of scientific papers that demonstrate the power of neuropsychological techniques in improving the lives of large numbers of people all over the world.[2]

Inspiration from Loved Ones with Brain Damage

Many people initially become interested in clinical psychology because they have a personal or family history of mental illness. They experienced and/or witnessed the tremendous personal toll that depression, anxiety, schizophrenia, and other conditions can take on a person, and they were inspired to help others who have suffered in a similar manner. The same is true for many neuropsychologists. For example, a neuropsychologist might have a family member who is unable to talk and walk after a severe traumatic brain injury. They might then observe the profound impairments that emerge in the acute phase post-injury, as well as the incredible degree of improvement and healing that takes place as a result of months of intensive cognitive and physical rehabilitation.

This personal or family experience is by no means required in order to pursue neuropsychology, but it does offer a first-hand perspective and a unique vantage point from which to understand brain disease and/or injury. For the subset of us who have been impacted by family neurological and psychiatric illnesses, this experience can allow for the development of a meaningful connection with other people who are in a similar circumstance. It can be an inspiration – a way to take this immense misfortune and struggle experienced by you, your family member, and/or your friend, and transform the stress and sadness into fuel and passion to help others who are faced with a similar burden. To better illustrate this point, both of us have life experiences we would like to share with you that influenced our decision to become neuropsychologists.

I (Ryan) have two close family members who experienced significant brain injuries. In 1984, when my Uncle Alan was in his mid-20s, his job was to drive and unload a frozen food truck. On one ill-fated day, he opened the back of the truck and a box fell from the top of a pallet and hit him directly on the head. A frozen box of this sort is not all that different from a large, heavy rock, and so it packs quite a punch. Still, somehow, he eventually stood up, dusted off, and went about his day. He did not go to the hospital and he continued to function for the next few weeks, although people who interacted with him said that he seemed a bit disoriented and confused during this time. As family legend has it, Thanksgiving Day arrived soon after the injury and he played in the annual family football game. At the very end of the game, with dinner ready and everyone waiting impatiently for the game to

---

[2] Listen to NavNeuro episodes 15 and 39 for discussions with Dr. Twamley and Dr. Smith, respectively.

end, Uncle Alan ran downfield, dove, and caught the game-winning touchdown. However, the whiplash from the fall may have aggravated intracranial (within the brain) swelling from his prior injury, and he did not get up from the catch. He was rushed to the hospital and he remained in a coma for several months afterward. Although he did eventually wake up, he has been severely brain-damaged and unable to care for himself for the past 30+ years. He cannot walk, he has very little mobility in his left arm, he has strabismus (one eye points up and to the side), and his cognitive abilities are severely impaired. My grandmother dedicated a huge portion of her life to Uncle Alan's care, initially in the home with her and my grandfather, and even after he was transferred to an assisted living facility. I have visited Uncle Alan countless times and his cutesy humor is always endearing (my long-time nickname has been "Cryin' Ryan"). I have also had the chance to interact with the neuropsychologist at Uncle Alan's assisted living facility over the years, and I have long speculated about what exactly happened to his brain on that day so many years ago. "How did he survive the injury?" "Why did the touchdown catch (but not the impact from the falling box) cause him to go into a coma for so long?" "What exactly happened to his brain such that he survived but lost so much in terms of his cognitive abilities?" Questions such as these have simultaneously saddened and fascinated me for years.

In another unfortunate event, my mother developed an arteriovenous malformation (AVM; a congenital tangle of blood vessels that is at risk of hemorrhage) in her brain. When I was five years old, we were living in central Virginia with my older brother. For no apparent reason, her AVM ruptured, causing her to bleed into her brain. She was airlifted to a nearby hospital and spent the next several months in a coma. I stayed with my aunt, uncle, and cousins in New York for the summer, waiting to find out what would happen to her. Ultimately, thankfully, she emerged from the coma and recovered enough to live on her own and care for me again. But she sustained significant brain damage and has had trouble with her vision, balance, and memory, as well as difficulties with stress and anxiety, ever since then. Although (unlike my uncle) people who meet her out and about at the grocery store or the mall often have no idea that there is anything wrong with her brain, she is constantly plagued by the impairments she sustained on that day nearly 30 years ago. I experienced her struggles firsthand as I grew up with her and I tried my best to help her as she tried her best to take care of me. I wondered, "How is her brain injury different from Uncle Alan's?" "Why does she look normal to most people while Uncle Alan does not?" "What caused both of them to emerge from their comas?" My questions are never-ending and both my uncle's and my mother's injuries led me to spend countless hours contemplating how the brain works. When I attended high school and college, I was drawn to subjects that offered insight into the brain, including neuroscience, biopsychology, and, ultimately, neuropsychology. I am eternally grateful to my uncle and my mother for the experiences that I shared with them, which helped me develop empathy for people with brain injuries, as well as curiosity about the inner workings of the brain. These experiences have shaped my thinking for years and ultimately led me down the path to becoming a neuropsychologist. In my work in this field, I hope that I can have a positive impact on the lives of many people, thereby transforming the pain and suffering of my family members into recovery and healing in others.

I (John) had the unfortunate experience of watching my grandmother slowly lose her ability to learn new information and form memories. It came on very gradually over the course of several years, first presenting as mild forgetfulness of conversations, then to repeating the same question five times in the course of a couple of minutes, to not recognizing her own husband, children, or grandchildren, to her current state of becoming very anxious that her parents (who have been dead for over 30 years) must be worried about her because she has not been home for several days. She is currently in the moderate stage of Alzheimer's disease, and I know what awaits her and my family in the coming years if she lives long enough: she will eventually be unable to remember who she is or where she is, will not be able to communicate with others, and will lose the ability to manage the most basic of activities, like feeding or dressing herself.

I use the experience with my grandmother to help others prepare for what might lie ahead. I encourage them to join research studies so that we might find a treatment for this awful disease that took my grandmother from me. I can empathize more with my patients' families and caregivers, knowing firsthand what it feels like to see a loved one's cognitive abilities slowly slip away. I have an extra drive to teach everyone that, although it is sometimes unavoidable, there are ways to reduce our own risk of cognitive decline through lifestyle factors such as exercise, healthy diet, sleep, psychological well-being, and strong social connections. And I have witnessed the fact that life can be beautiful and worthwhile in any stage or condition. I see the joy that my family and I experience just from being around my grandmother, even if she does not know who we are or what year it is.

Developing Brain-Behavior Expertise

If the human brain were so simple that we could understand it, we would be so simple that we could not.

– Emerson W. Pugh, as quoted in the 1977 book *The Biological Origin of Human Values*

It never ceases to amaze us that the three-pound bag of tissue, chemicals, and electricity in our skulls is the most complex structure in the known universe. The number of neurons (~86 billion; Azevedo et al. 2009) is close to the number of stars in the average galaxy. The number of total connections between different brain cells is exponentially higher and likely uncountable. This is why Emerson Pugh's quote is so on-point. If you want simple and vanilla, then neuropsychology is the wrong profession for you. However, if you are seeking a lifetime of fascination, investigation, and mind/brain knowledge, then you're in the right place.

Because we aspire to be brain-behavior experts, our work takes us directly to the core of the one organ that truly defines who we are as human beings. The brain is the basis for love and hate, passion and boredom, courage and fear, pleasure and pain, hunger and satiety, and all of the other subjective and unconscious experiences that define our lives. Moreover, there is nothing more intimate to us than our cognitive abilities. No matter what changes occur in the world around us, we rely on our ability to think, to communicate with other people, and to solve problems. This makes us who we are as individuals and as a species. In his 2011 book, *Sapiens,* historian and author Yuval Noah Harari discusses the importance of the cognitive revolution – an evolution of the cerebral cortex that is estimated to have been in effect approximately 70,000 years ago – in shaping humans as we are today. Indeed, it is our cognitive abilities that distinguish us from other animals (who, by the way, can be very intelligent in their own right) and allow us to accomplish incredible feats such as communicating with friends and family on other continents, building 2700-foot skyscrapers, and even traversing the solar system. On an individual level, our brains have undergone an unbelievable transformation over the course of our lives, from a tiny ball of cells in our mother's belly (the *neural tube*) to a structure that powers an autonomous, fully functional person with complex thoughts, emotions, and values. Neuropsychologists are in the business of understanding the thoughts and behaviors that arise from the brain and helping people maintain these abilities for as long as possible. We think that this makes our field both benevolent and scientifically cutting edge.

> I want to feel intellectually engaged in my daily activities, and the study of brain-behavior relationships readily allows for this. As a relatively nascent science, neuropsychology has a high rate of knowledge turnover. As such, the opportunity to meaningfully contribute to advances in our understanding of brain-behavior relationships is highly appealing and exciting.
>
> – Charles Gaudet, PhD

As you may be aware, the brain is often touted as the new frontier of scientific inquiry and understanding. Although it has been studied since at least the time of Ancient Egypt, it remains largely a mystery. Said another way, if we put together all that we have learned about the brain in the last few thousand years and stack that up next to all that we do *not* yet know about the brain, there is no comparison. What we do not know dwarfs what we do know. Still, there is reason to be optimistic. We currently have tools at our disposal that would have probably been considered somewhere between science fiction and magic as recently as a century or two ago – tools such as functional magnetic resonance imaging (fMRI) and positron emission tomography (PET) that allow us to peer inside the skull of a living person in ways we never thought possible. Largely because of these technological advances, as well as an increase in funding for research, we have learned more about the brain in the past few decades than in the entirety of premodern human history. And we have reason to believe that this progression will continue, likely at even more rapid rates. For example, the White House Brain Research through Advancing Innovative Neurotechnologies (BRAIN) Initiative announced by the Obama Administration in 2013 is a collaborative research enterprise aimed at revolutionizing our understanding of the human brain, and it has already garnered awards totaling approximately $1.3 billion through 2019, per their website (https://www.braininitiative.nih.gov/about/overview).

There are high hopes that, by the end of this century, we will have developed a type of precision medicine that would allow us to selectively alter aberrant brain processes in ways that would prevent or cure mental illnesses and brain-related diseases. If everything reduces down to electrochemical processes, then we should in principle be able to precisely change anything we want about the brain and behavior. For example, theoretically, it should be possible to "download" a new language or mastery of jujitsu, similar to what is seen in the movie, *The Matrix*. As an example of recent advances, new gene-editing techniques (e.g., CRISPR-Cas9) hold promise for making fundamental, long-lasting changes to human DNA. In addition, even brain and spinal cord injuries that currently leave people without the ability to speak or move will ultimately be readily treated, and researchers have already begun to translate neural activity into speech and actions (e.g., Anumanchipalli et al. 2019; Grahn et al. 2014). The more we understand about the brain, the more we will understand consciousness and the better we will be at creating artificial general intelligence (Tegmark 2017). The potential for discovery and innovation is endless, and neuropsychology is intricately tied to both the process and the outcomes of this brain science research.

A Melting Pot of Knowledge and Skills

Neuropsychology is an amalgam of different disciplines. The term is itself a portmanteau of the words "neuro" (i.e., neuron, or the study of neuroscience) and "psychology," (i.e., the study of behavior and thinking), but it also incorporates additional aspects of medicine, statistics/psychometrics, and philosophy, among others (see the Fig. 2.1). We think that this is one of the most unique and attractive aspects of our field – that we draw upon so many different areas of knowledge on a daily basis.

> My family has some background in medicine, so I was interested in medicine, particularly within oncology, and started thinking a little bit about that, but realized that I didn't care so much for biology or chemistry and medical school was probably not going to be the best choice for me. In college, I had started studying memory with one of my professors and was very intrigued by that… and that really sort of spurred me on to look more into it. Again, my path wasn't direct, so I started off working more with adults in behavioral medicine related to oncology and then realized that I enjoyed working more with children, moved into child clinical psychology and realized that I still wanted that medical piece to come back and found neuropsychology to be the perfect blend of all of my areas of interest.
>
> – Christine Trask, PhD, ABPP-CN

Similar to Dr. Trask, many neuropsychologists are initially very interested in the medical field; this makes sense because there is a large amount of overlap between neuropsychology and human neurobiology. In our training, we spend a significant amount of time learning about neuroanatomy, neuroscience, and medical conditions

Fig. 2.1 The interdisciplinary "melting pot"

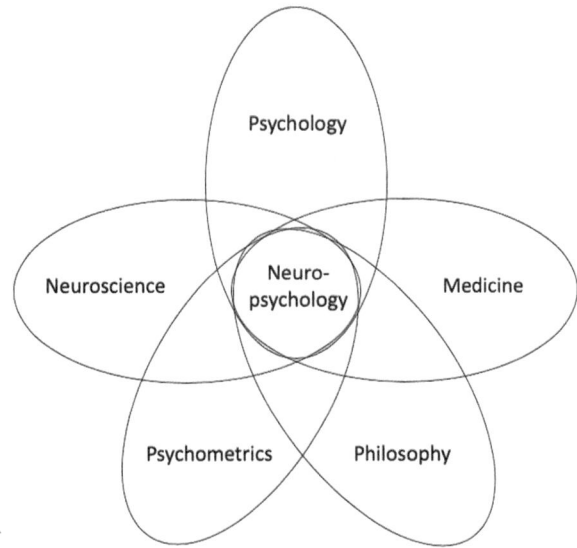

with the potential to impact our cognitive abilities. Many of us find this aspect of our training to be particularly fascinating. From killer T immune cells mounting attacks on invading bacterial infections to cellular mitochondria working like miniature factories, pumping out chemicals and proteins, our bodies are incredible, complex machines. Many scientists believe that our brain is the most impressive component of this machine. The basis of our intellect, consciousness, and behavior is built upon simple mechanisms called action potentials, where an electrical signal is sent from one cell to another across a small gap called a synapse. The manner in which trillions and trillions of action potentials work together in synchrony to produce our conscious experience is called the "hard problem of consciousness" and is still a matter of great scientific and philosophical interest and investigation (Chalmers 1996; Harris 2019).

> My journey to becoming a clinical neuropsychologist began when I was a pre-medical undergraduate student, majoring in Neuroscience with a minor in Philosophy. My immersion into two seemingly contradictory yet remarkably analogous fields of study shaped my fascination with understanding the neurological bases of human experience. On one hand, I studied the intricacies of the brain at the molecular, cellular, and systems levels; on the other, I sought to resolve the 'mind-body problem' posed by Plato and Descartes and grasp the relationship between the brain and human consciousness. I was drawn to neuropsychology, as it allowed the opportunity to ask questions regarding human behavior and examine psychological phenomena through biological processes – perfectly merging my passions for science and the philosophical notion of what it means to be human.
>
> – Tanya Nguyen, PhD

Dr. Nguyen's reason for entering the field is very similar to my (John) own. In high school, I became interested in existential and epistemological philosophy (that's a whole separate story!) and had been considering pursuing a career as a philosophy professor. However, in my junior year of college I took a biopsychology class that lit a spark in me. The professor focused on the philosophical aspects of neuroscience such as how little we understand about consciousness (e.g., the hard problem of consciousness mentioned above), the question of whether or not we have free will, and the search for the engram (how memories are physically stored). The prospect that I could continue contemplating these deep philosophical questions while also studying the brain, contributing to scientific knowledge, making a profound impact in people's lives, and earning a decent salary while doing it… it was a no-*brain*er!

> When I started in my clinical psychology doctoral program, I aspired to be a therapist. As I began my training, I enjoyed working with patients, but didn't feel entirely fulfilled by the practice of psychotherapy. As part of my training, we were required to take some classes related to neuropsychology and to complete some neuropsychological test batteries in our practica. I had found my home! …I've never regretted my decision to pursue neuropsychology and I encourage students considering psychology to seriously consider neuropsychology as part of their own career path.
>
> – Megan Spencer, PhD

Psychology itself is a vast ocean of subspecialties and different areas of expertise. As psychologists, we gain a large breadth of knowledge and training. For example, in addition to all of our coursework on child development, statistics, health psychology, psychopathology, psychopharmacology, etc., we also engage in novel research, conduct psychotherapy sessions with patients, and learn how to teach and mentor others. This breadth of training lays the foundation from which we are then able to specialize in assessing cognition. It takes nothing less than all of the skills molded in graduate school to do the work of a neuropsychologist.

****

While three activities (clinical work, research, and teaching) are how most of us spend the majority of our time, there are many other potential roles and ways to divide one's working hours. For example, we can spend time educating the general public about brain health, participating in local government to help children with developmental delays, joining a professional committee to improve student outreach, hosting a podcast, or even writing a book ☺.

Novelty and Intellectual Engagement

One question that everyone should ask themselves prior to embarking on a long path toward specialization in a field is, "After all of this education and training (and tuition costs), will I just end up being bored after a few years on the job?" We have yet to find a colleague who would endorse that sentiment.

> …I caught the bug and still am amazed and grateful to be part of such an exciting, dynamic, and rewarding field.
>
> – Nancy Nussbaum, PhD, ABPP-CN

> This is a great field to go into. It has never gotten old, I've never gotten bored with it, and it is going to be exciting for years to come.
>
> – Peter Arnett, PhD

> I can honestly say there has never been one dull day in my career as a UCLA Neuropsychologist. I never tire of the analysis of the data, the combing through research, the delivery of the results to patients/families, the creation of cognitive rehabilitation programs, and the desire/need to share it all with a fresh crop of graduate students each year.
>
> – Karen Miller, PhD

Dr. Miller's enthusiasm is a testament to her personality, but it also reflects the dynamic nature of the field. In neuropsychology, you never stop learning and acquiring new knowledge and skills. Neuropsychologists receive foundational training in graduate school and advanced education during the internship and postdoctoral years, but in some sense that is just the beginning of the learning process. Throughout our entire careers we are exposed to patients with symptom presentations that we have never seen before, we embark on research projects to answer questions that have never before been answered, and we take on new challenges that lead us to push ourselves in ways we never thought possible. This can feel overwhelming at times, but we love the fact that there is always something new to discover and that novel scientific findings enhance our understanding of the connection between the brain and behavior on a daily basis.

When I discovered clinical neuropsychology – a career that would allow me to help people by solving the puzzles of how their brains function – it was a perfect fit!

– Taylor Greif, PhD

Conducting a clinical neuropsychological evaluation can feel like working on a complex puzzle. First, we put together a patient's reported clinical history and medical record information with an eye for pertinent diagnostic and prognostic information. Next, we incorporate cognitive test scores, including strengths and weaknesses across a variety of specific abilities, and we pull out the pieces of the presentation that fit together to explain the patient's symptoms. Let's walk through an example. First, an older male patient may tell us that his bed partner has complained that he moves and thrashes about in his sleep. He might also talk about feeling unsteady on his feet and feeling less motivated to get out of the house and go golfing with his buddies. He might then describe changes in his handwriting when he writes checks such that it has become much smaller and harder to read. He could also throw us a curveball and talk about unusual sensory experiences, where he is frightened by the sight of mice running around under his kitchen table, even though his wife insists that there are no mice in the kitchen. His neuropsychological test scores could show average memory abilities with impairments in visuospatial skills, attention, and executive functioning. This may all sound like random bits of information to a layperson, but a neuropsychologist would process each of these puzzle pieces and then thoughtfully put them together into a probable diagnosis of Lewy body dementia. If this term sounds familiar, you may have heard about it on the news because the actor Robin Williams was diagnosed with this condition prior to his tragic death.

To add to the Lewy body dementia example, a few more excellent analogies were drawn by Leslie Gaynor (doctoral student in neuropsychology at the University of Florida) when she said that a neuropsychologist is *"like a detective, a technician, a scientist, a clinician, and a storyteller, all in one."* From this perspective, each patient can feel like a new episode of Law and Order or Crime Scene Investigation (CSI), and the neuropsychologist is the detective whose job it is to solve the mystery of what is causing/contributing to the cognitive impairments in their patient. This involves frequent critical thinking and problem solving and it constantly drives us back to the scientific literature for guidance.

Prestige

Not surprisingly, we did not collect any quotes from colleagues saying that prestige was the primary reason why they chose to pursue neuropsychology. Still, it is natural to value status and respect, and it doesn't necessarily signal narcissism or selfishness. Indeed, the drive to succeed and earn social status can be leveraged to do a lot of good in our world, and it is probably at least part of the motivation behind many famous social activists, political leaders, and scientists.

In order to become a neuropsychologist, you first earn a doctorate (PhD, PsyD, or, less commonly, EdD), which is a great honor in and of itself. What's more, in our experience, when people ask us what we do for work (e.g., at a cocktail party or the barbershop) and we say, "neuropsychologist," and then describe our jobs, it often leads to interesting and engaging conversations. Many people find the brain to be fascinating and, unlike some scientists who study niche topics that are less relatable, the study of cognition is relevant to every single person on the planet. Therefore, our job titles, and the knowledge we gain while attaining them, make us experts in cognitive abilities and also increase our status. Personally, we derive a great deal of pride and sense of accomplishment from being good at what we do and being asked by other professionals to help them better understand their patients.

Financial Compensation

You will also probably never hear someone say that they pursued neuropsychology primarily for the money because there are *much* easier ways to earn an above-average income that do not involve 4 years of college, about 6 years of graduate school, and then 2 years of fellowship. However, it is possible to make a comfortable living (well above the median US household income) and I (John) will be honest when I say that it factored into my decision to pursue this field. Among psychologists, we are one of the most well-compensated specialties (Sweet et al., 2020), and our job flexibility allows us the choice to pursue niche areas within neuropsychology that are more financially lucrative (i.e., legal/forensic work).

There is a Salary Survey published about every 5 years that polls a large sample of neuropsychologists to inquire about their annual income and other job characteristics (Sweet et al. 2020). Per the 2020 survey, median annual income during the first 5 years after the postdoctoral fellowship is $110,000 and increases to $126,000 by years 6–10. As you would expect, there is a steady increase in income with years of experience. For example, the median annual income is $146,500 for those with 11–15 years of experience and $220,000 for those with >25 years of experience. Importantly, the data vary somewhat depending on the geographic region. For example, the median income in California is $150,000, compared to $117,000 in North Carolina. And, due to economic inflation, the values we just reported will likely underrepresent the salaries of neuropsychologists in the coming years. Finally, the job setting significantly impacts income as well, with clinicians in private practice settings reporting the highest earnings (possibly because these professionals see the most legal/forensic cases), and neuropsychologists in university psychology departments earning the lowest salaries.

Although the aforementioned averages are lower than other highly compensated professions (e.g., physician, hedge fund manager, attorney), they are higher than most other nonphysician "helping professions" and can easily make for a comfortable life. This means that you do not have to choose between your passion and your paycheck – you can have both![3]

Job Security

> I am pleased that I have had the chance to choose my own path, to contribute to an important field of science, to have 20+ years of working with patients and families, to help lead a department, and, most importantly, to have participated in the growth and training of many outstanding young psychologists and neuropsychologists.
>
> – Jeff Wozniak, PhD, LP

There is a huge societal need for neuropsychologists, and this gives us excellent job security. In the Salary Survey referenced above, the vast majority of respondents reported working full-time. Still, our jobs are vulnerable to macroeconomic effects, such as the recession caused by the COVID-19 pandemic. We do not have data on the number of neuropsychologists who were laid off and/or lost income due to the virus, but we do know that, in response, the field pivoted toward greater adoption of teleneuropsychology services, which further increased the robustness of our jobs in the long term.

Clinically, there are many people who need services, especially individuals of low socioeconomic status, people residing in rural communities, and people in developing nations. These groups tend to experience greater cognitive impairments than others due to poor nutrition, less access to high-quality healthcare, exposure to pollutants/ toxins, and other important factors. Neuropsychologists can help by leading the charge toward important scientific advances (e.g., What chemicals cause cognitive impairments? How do we detect early memory loss? How do we stave off and/or treat brain insults?) and by establishing clinics/practices to evaluate and treat them. Much of this work is embedded in the movement toward preventative healthcare, where we take action to prevent illness in the "preclinical" phase of a disease (i.e., before symptoms are present) rather than waiting until someone is already afflicted and then attempting to treat them. For example, many researchers attempting to find a cure for Alzheimer's disease are examining people who do not yet have any memory loss but whose brains are vulnerable to later developing dementia of the Alzheimer's type. Fortunately, governments and insurance companies have begun to realize that it makes more sense to spend a small amount of money to prevent chronic illnesses as opposed to spending a much larger sum to treat disease symptoms. Thus, more efforts are being focused on improving brain health and reducing risk of cognitive decline, and this has increased the demand for neuropsychologists all over the country.

Relevant to the topic of Alzheimer's disease, an important group of people who are in great need of help from neuropsychologists is older adults. As you may be aware, the population is rapidly aging. By 2050, 20% of people in the US are projected to be

---

[3] Despite our optimistic stance, student loan debt can be substantial. As such, we discuss finances in each chapter of Part II.

over age 65, with the largest growing segment of the population being 85 or older (US Census 2010). These demographic changes will bring with them an increase in age-related illnesses such as stroke, epilepsy, traumatic brain injury, and various causes of dementia. It is not an overstatement to label this phenomenon as a public health crisis, as it will put further stress on an already strained healthcare system and will greatly increase the emotional, physical, and financial burden on caregivers. Neuropsychologists are indispensable members of healthcare teams who evaluate, manage, and treat older adults with cognitive impairments, and we will continue to provide outstanding care for older adults in the years to come.

On the other side of the age spectrum, rates of pediatric disorders such as ADHD, learning disabilities, and autism spectrum disorder continue to increase as well (Boyle et al. 2011; Visser et al. 2010). The vast majority of parents have their children's best interests in mind and they want to have their children evaluated for these conditions by neuropsychologists, so waitlists for these providers tend to be quite long. As the rates of these disorders continue to climb, so will the need for clinicians and researchers.

In the *Melting pot of knowledge and skills* section, we discussed the flexibility of the profession, and this flexibility improves job security because our degree confers us the ability to wear many different hats. For example, in the unlikely event that our employer (e.g., a hospital) determines that they are no longer able to support our salary and we are unable to find grant money to sustain our position, we have other options. We can teach courses at our local college, we can open a private practice, or we can take a job in industry (i.e., the private sector). In other words, the skills that we have developed (e.g., knowledge of functional neuroanatomy, differential diagnosis, research design, critical thinking, test administration, technical writing, intervention) are highly transferrable. This fact became evident for many neuropsychologists during the COVID-19 pandemic, when they temporarily stopped seeing patients and testing participants in person and transitioned to using their other skills while clinics and research labs were shut down.

Even outside of a pandemic, many institutions have flexibility built into the position such that we are not forced to work the traditional 9–5 shifts while tied to a desk in a small cubicle every single day. As a few examples, many settings allow for a compressed workweek (e.g., four 10-hour days), working from home, and working while traveling to conferences and meetings. Many young parents (us included) find this flexibility to be incredibly helpful, as they are able to tailor their schedule to childcare arrangements.

Conclusion

As we have repeated multiple times in this chapter, most neuropsychologists find *many* significant advantages to working in this field. We hope that at least some of the reasons outlined in these pages have sounded appealing to you as well. Remember that no matter what path you ultimately decide to pursue, it is vital that you develop your own personal, "Why?" With that in mind, we truly hope that our love for neuropsychology came across in these pages and that you will seriously consider this career because, as Dr. Postal said, we think that we are in the best field ever.

References

Allott, K., Brewer, W., McGorry, P. D., & Proffitt, T. (2011). Referrers' perceived utility and outcomes of clinical neuropsychological assessment in an adolescent and young adult public mental health service. *Australian Psychologist, 46*, 15–24.

Anumanchipalli, G. K., Chartier, J., & Chang, E. F. (2019). Speech synthesis from neural decoding of spoken sentences. *Nature, 568*(7753), 493–498.

Ayers, C. R., Dozier, M. E., Twamley, E. W., Saxena, S., Granholm, E., Mayes, T. L., & Wetherell, J. L. (2018). Cognitive Rehabilitation and Exposure/Sorting Therapy (CREST) for hoarding disorder in older adults: A randomized clinical trial. *The Journal of Clinical Psychiatry, 79*(2), 16–22.

Azevedo, F. A., Carvalho, L. R., Grinberg, L. T., Farfel, J. M., Ferretti, R. E., Leite, R. E., et al. (2009). Equal numbers of neuronal and nonneuronal cells make the human brain an isometrically scaled-up primate brain. *Journal of Comparative Neurology, 513*(5), 532–541.

Boyle, C. A., Boulet, S., Schieve, L. A., Cohen, R. A., Blumberg, S. J., Yeargin-Allsopp, M., et al. (2011). Trends in the prevalence of developmental disabilities in US children, 1997–2008. *Pediatrics, 127*(6), 1034–1042.

Chalmers, D. J. (1996). *The conscious mind: In search of a fundamental theory*. Oxford University Press.

Grahn, P. J., Mallory, G. W., Berry, B. M., Hachmann, J. T., Lobel, D. A., & Lujan, J. L. (2014). Restoration of motor function following spinal cord injury via optimal control of intraspinal microstimulation: Toward a next generation closed-loop neural prosthesis. *Frontiers in Neuroscience, 8*, 296.

Greenaway, M. C., Hanna, S. M., Lepore, S. W., & Smith, G. E. (2008). A behavioral rehabilitation intervention for amnestic mild cognitive impairment. *American Journal of Alzheimer's Disease & Other Dementias, 23*(5), 451–461.

Harari, Y. N. (2011). *Sapiens: A brief history of humankind*. Vintage Books.

Harris, A. (2019). *Conscious: A brief guide to the fundamental mystery of the mind*. HarperCollins Publishers.

Hilsabeck, R. C., Hietpas, T. L., & McCoy, K. J. M. (2014). Satisfaction of referring providers with neuropsychological services within a veterans administration medical center. *Archives of Clinical Neuropsychology, 29*, 131–140.

Mahoney, J. J., Bajo, S. D., De Marco, A. P., Arredondo, B. C., Hilsabeck, R. C., & Broshek, D. K. (2017). Referring providers' preferences and satisfaction with neuropsychological services. *Archives of Clinical Neuropsychology, 32*, 427–436.

McCrae, M. (2008). *Mild traumatic brain injury and postconcussion syndrome: The new evidence base for diagnosis and treatment*. Oxford University Press.

Medalia, A., & Erlich, M. (2017). Why cognitive health matters. *American Journal of Public Health, 107*(1), 45–47.

Rosado, D. L., Buehler, S., Botbol-Berman, E., Feigon, M., León, A., Luu, H., et al. (2017). Neuropsychological feedback services improve quality of life and social adjustment. *The Clinical Neuropsychologist, 32*(3), 422–435.

Stark, D., Thomas, S., Dawson, D., Talbot, E., Bennett, E., & Starza-Smith, A. (2014). Paediatric neuropsychological assessment: An analysis of parents' perspectives. *Social Care and Neurodisability, 5*, 41–50.

Sweet, J. J., Klipfel, K. M., Nelson, N. W., & Moberg, P. J. (2020). Professional practices, beliefs, and incomes of US neuropsychologists: The AACN, NAN, SCN 2020 practice and "salary survey". *The Clinical Neuropsychologist*, 1–74.

Tegmark, M. (2017). *Life 3.0: Begin human in the age of artificial intelligence*. Vintage Books.

Temple, R. O., Carvalho, J., & Tremont, G. (2006). A national survey of physicians' use of and satisfaction with neuropsychological services. *Archives of Clinical Neuropsychology, 21*(5), 371–382.

Twamley, E. W., Thomas, K. R., Burton, C. Z., Vella, L., Jeste, D. V., Heaton, R. K., & McGurk, S. R. (2017). Compensatory cognitive training for people with severe mental illnesses in supported employment: A randomized controlled trial. *Schizophrenia Research, 203*, 41–48.

Twamley, E. W., Vella, L., Burton, C. Z., Heaton, R. K., & Jeste, D. V. (2012). Compensatory cognitive training for psychosis: Effects in a randomized controlled trial. *The Journal of Clinical Psychiatry, 73*(9), 1212–1219.

U.S. Census Bureau. (2010). *The older population: 2010*. https://www.census.gov/prod/cen2010/briefs/c2010br-09.pdf

VanKirk, K. K., Horner, M. D., Turner, T. H., Dismuke, C. E., & Muzzy, W. (2013). Hospital service utilization is reduced following neuropsychological evaluation in a sample of US veterans. *The Clinical Neuropsychologist, 27*(5), 750–761.

Visser, S. N., Bitsko, R. H., Danielson, M. L., Perou, R., & Blumberg, S. J. (2010). Increasing prevalence of parent-reported attention-deficit/hyperactivity disorder among children – United States, 2003 and 2007. *Morbidity and Mortality Weekly Report, 59*(44), 1439–1443.

Watt, S., & Crowe, S. F. (2018). Examining the beneficial effect of neuropsychological assessment on adult patient outcomes: A systematic review. *The Clinical Neuropsychologist, 32*(3), 368–390.

Chapter 3

Where Do Neuropsychologists Work?

> Neuropsychologists can contribute their expertise in so many different clinical and research settings... if variety is truly the spice of life, the field of neuropsychology is one of the spiciest.
>
> – Adam Parks, PhD, ABPP-CN

In the previous chapter, we presented a variety of reasons for pursuing a career in our field. If your interest in neuropsychology has been piqued, you might be wondering, "Where do I find neuropsychologists out there in the world?" and "If I choose to pursue this field, where would I spend most of my time?" The purpose of this chapter is to answer these questions in detail.

The following sections describe the most common settings where neuropsychologists work. We will elaborate on unique facets of the environments we cover, but it is important to note that (a) these institutions are frequently linked together (e.g., a university-affiliated hospital) and (b) there is significant overlap across different sites – for example, many characteristics of rehabilitation hospitals also apply to freestanding hospitals. There is also variability depending on the specific geographic location (e.g., the VA medical center in Albuquerque, NM, looks different from the VA hospital in West Haven, CT), and there are rare exceptions to most of what we cover. Overall, the goal of this chapter is to provide you with a sense as to what the world around you might look like if you pursue a career in neuropsychology.

© Springer Nature Switzerland AG 2021
J. A. Bellone, R. Van Patten, *Becoming a Neuropsychologist*,
https://doi.org/10.1007/978-3-030-63174-1_3

Universities

If you have already started or finished college (which is by no means a prerequisite for reading this book), you are probably familiar with what a typical university looks and feels like. The term "university" comes from the Latin, "universitas magistrorum et scholarium," which can be translated to "community of teachers and scholars." Many people consider the Greek philosopher Plato to have founded the first university when he opened his famous *Academy* in circa 387 BCE. His student Aristotle was later trained there and then went on to establish the foundation of our knowledge in ethics, biology, psychology, and politics. Fast forward to the twenty-first century and there are currently over 5000 colleges and universities in the United States and over 20,000 worldwide. These environments are proverbial hotspots of new learning and discovery, and many neuropsychologists are drawn to them like flies to honey.

But what do neuropsychologists spend their time doing at universities? Neuropsychologists are initially trained as psychologists; consequently, they are typically housed in a Department of Psychology, alongside other subspecialists such as social psychologists, cognitive psychologists, developmental psychologists, and industrial-organizational psychologists. Alternatively, if the university is affiliated with a medical center (see *Academic medical center*, below), a neuropsychologist could be housed in a Department of Neurology or Psychiatry, in which case they are more likely to interact frequently with medical doctors. In either case, our typical responsibilities include teaching courses, mentoring students, leading research labs, participating in university-based committees, and completing other administrative work.

There are three primary classes of positions available to all university professors, including neuropsychologists: tenure track, grant funded, and adjunct (see Table 3.1). Tenure-track faculty positions are very popular, and so they are quite competitive and often challenging to find. People are attracted to these openings for a number of reasons, including a high level of intellectual stimulation, the ability to mentor students, excellent job security, and a regular salary and benefits. The process of applying for and receiving tenure can be drawn out and demanding, but once a faculty member is "tenured," they have great academic freedom and flexibility to

Table 3.1 University professorships

Position	Main duties	Stability	Salary/benefits
Tenure track	Teach courses, conduct research, perform administrative work	High stability, especially once tenured	Medium
Grant funded	Conduct research, supervise students, teach courses, perform administrative work	Moderate stability; job is reliant upon grant funding	High ceiling but dependent upon grants
Adjunct	Teach courses	Low stability (but high flexibility)	Low

pursue their research and teaching interests (i.e., they have fewer mandatory productivity demands, and they are rarely fired). Tenure-track faculty members often spend a significant proportion of their time teaching and conducting research, and they also perform administrative tasks such as attending faculty meetings and reviewing applications from prospective students.

An alternative to a tenure-track professor is a grant-funded researcher. In a tenure track, the person earns "hard money" (i.e., a consistent, stable salary), while grant-funded faculty make "soft money" (i.e., the more grants you are awarded, the greater your income).[1] One way to conceptualize a soft money job is as similar to a primarily or exclusively commission-based sales position – the more you sell, the larger your bank account. An advantage of soft money positions is that they tend to have higher financial ceilings; that is, your maximum income is greater than in a hard money job. But you may have guessed the flip side to this coin – if you are not awarded grants, your income is not guaranteed. It is important to note that it is possible to blend these hard and soft positions such that a portion of one's salary is guaranteed and the other portion is grant-based, but it is also common to have a 100% soft or hard money job. As we discussed in the previous chapter, neuropsychologists also have other available options with which to supplement their income, including teaching additional courses and picking up clinical work. Overall, if you are very passionate about research and you want maximal flexibility and creative license in designing your own research studies, we encourage you to consider a university-based, grant-funded position. These people dedicate more time than anyone to pursuing novel, creative, innovative studies that have the potential to alter and improve our field for decades to come.[2] Alternatively, if occupational security, financial stability, and teaching/education are paramount to your job satisfaction, then a tenure-track faculty position may be a better fit for you.

In contrast to tenure-track faculty and grant-funded positions, the final primary faculty designation is as an adjunct professor. You can think about this role as broadly similar to a part-time job at a retail store, restaurant, or office setting. These jobs tend to pay less, to have fewer fringe benefits, and to have far less stability than the other two positions. However, there are upsides to adjuncting as well, and we have both done it ourselves. Specifically, John taught a *Psychology of Aging* course at Providence College in Rhode Island during his postdoctoral fellowship at Brown University and Ryan taught a *Clinical Psychological Science: Theory & Methods* course at San Diego State University during his postdoctoral fellowship at UCSD. The adjunct professorship is an excellent way to gain teaching experience early in your career (which increases your competitiveness for those sought-after tenure-track faculty positions) while also earning a bit of pocket change, all without requiring a long-term commitment or the

---

[1] Neuropsychologists frequently submit grant applications to the National Institute of Health; https://www.nih.gov/

[2] A lot goes into writing a grant. You are trying to sell yourself, your research team, your institution, and, most importantly, your research idea. This topic is beyond the scope of our book, but we can provide a few solid resources: Gajda and Tulikangas (2005); Zlowodzki et al. (2007); Robertson, Russell, & Morrison (2020).

burden of administrative tasks. Many people choose to do what we did and supplement their current jobs with an adjunct position.

****

It is probably clear by now that, overlapping with the three classes of professorships, there are also two central roles that neuropsychologists and other academics take on: teacher and researcher. Let's dive deeper into these concepts, beginning with the former. Teaching can be an incredibly rewarding and fulfilling job. I (Ryan) know that I would not be where I am today without some of the mentors with whom I interacted along the way. I completed my undergraduate degree at James Madison University (JMU) in Harrisonburg, VA, and I was lucky enough to enroll in a statistics course with Dr. Kevin Apple in my first semester. I then proceeded to work with him for 3 years, and we still share emails and the occasional breakfast or lunch when I am in town. His warm, caring demeanor, his authentic interest in my studies and my life in general, and his witty humor have all stayed with me to this day. He taught me countless foundational concepts and facts in psychology, and I enjoyed every minute of it. I recall arriving to my first exam in his statistics course, opening the door, and being astonished to hear the song *Eye of the Tiger* by Survivor (the *Rocky* song) playing on the speakers. It was one of many strategies he used to help move us into "the zone" (the *flow state* for you nerds out there) leading up to the test. For a group of apprehensive college students, it was truly music to our ears, and it helped us to decompress and share a smile before the big exam. I could provide many more anecdotes about Kevin Apple, but I will spare you the tangents. My point is that a teacher shaped my knowledge and thinking styles from an early stage of my career, placing me on the path to where I am today, and I remain eternally grateful. I have used Dr. Apple as an exemplary teacher here, but John and I have both been lucky enough to work with dozens of other wonderful mentors, supervisors, and instructors as well, from high school through postdoctoral fellowship.

Not surprisingly, many of our own teachers have been neuropsychologists. People in our profession teach a variety of courses, ranging from statistics to psychology 101 to psychopathology to psychopharmacology. However, we are most commonly sought after as instructors for courses in psychological assessment, functional neuroanatomy, neuroscience, brain and cognitive disorders, and lifespan cognitive development. These topics are our bread and butter.

Importantly, teaching involves much more than simply standing at the front of the class and lecturing to students for hours on end. On the contrary, we think about teaching as a holistic, inclusive, experiential activity, the core of which involves guiding a less experienced person on a path of greater knowledge, skills, and wisdom. There can be immense, long-lasting benefits to this activity, both for the teacher and for the learner.

In addition to teaching courses, university professors also frequently spend a large proportion of their time mentoring students. This includes one-on-one discussions about the students' academic progress, career goals, research interests, and other important topics. It also includes a large amount of time spent in small groups, often as part of the professor's research lab. One of the truly rewarding aspects of experiences such as these is that the mentor has the pleasure of witnessing generations of students advancing from novice to expert. Mentors observe their former students transition to other institutions and conduct their own research while taking pride in the fact that they are instrumental in helping their students develop the skills necessary to make discoveries of their own. As they supervise further generations of budding professionals, mentors continue to develop their "trainee family tree" and watch as their influence in the field increases exponentially.

****

As noted earlier, the other common aspect of a professor's role at a university is engagement in research. We already touched upon this topic when we described grant-funded positions, but we want to add a bit more information here. First, it is common for professors to develop and manage a "research program" or area of specialization at a university. Neuropsychologists have a range of brain-behavior research interests, from defining the effects of environmental toxins on reading proficiency, to programming and testing innovative computerized tests measuring attention and memory, to designing interventions for people caring for loved ones with dementia, and everything in between. Additionally, the research tools used by neuropsychologists differ from those of other professors. Specifically, most of us incorporate some form of cognitive and psychological testing into our research methods. For example, we might administer a neuropsychological test battery to patients before and after they receive a new cancer treatment in order to determine the extent to which it affects their thinking skills. Other potential tools in our tool belts include differential diagnostic skills (i.e., the ability to distinguish between different mental health disorders and brain pathologies) and cognitive training interventions to improve memory and problem-solving. One particularly cutting edge technique adopted by some neuropsychologists is neuroimaging, where we use functional magnetic resonance imagining (MRI) and/or other modalities to examine brain activity in response to tests of attention, working memory, and executive functioning (Bigler 2019).

Once a researcher develops expertise in a specific niche area, they are frequently asked to deliver presentations on the topic at conferences and other venues and/or to write scientific manuscripts, books, and book chapters on the subject. As a researcher, you can become an expert and the "go-to" person in your area(s) of specialty. You might be interviewed by journalists, reporters, or podcasters, and other professors and laypeople will frequently ask you questions about your area of expertise. This can be one of the most fun, engaging, and rewarding components of a research career.

An important aspect of all research in universities is mutually beneficial collaborations with other scientists, both those inside of and outside of your own discipline. In relationships with researchers in other fields, neuropsychologists are viewed as the experts in cognition – particularly in cognitive abilities in people with neuropsychiatric, disorders or injuries. Depending on the neuropsychologist's area of interest, he or she might collaborate with experts in internal medicine, cardiology, neuroradiology, neurosurgery, geriatric psychiatry, virology, computer science, or child development. In fact, it is difficult to think of a discipline that would not potentially need a brain-behavior expert for some aspect of a research project.

Most neuropsychologists work with human participants, but, interestingly, some of us conduct animal research as well. The brains of rats, mice, and primates are remarkably similar to human brains in a variety of ways, and we can learn a great deal of information by studying the neural functioning of these organisms. For example, in graduate school, I (John) collaborated with an electrophysiologist (a scientist who studies the electrical properties of cells) who received a large grant from the National Aeronautics and Space Administration (NASA) to study the impact of radiation exposure on mice with a predisposition to developing Alzheimer's disease-like brain pathology. Specifically, NASA administrators wanted to know whether or not the solar and cosmic radiation that astronauts would be exposed to on long space flights could possibly impair their cognitive abilities and jeopardize their ability to complete a mission and return home safely. My primary role in this project was to test the memory of mice before and after exposure to high-energy particles. This led to several publications, many conference presentations, and some cool stories (e.g., I proposed to my now-wife on a NASA-sponsored trip to Japan!).

****

In addition to teaching and research, there are also several other opportunities available at universities. Programs with master's and doctoral-level students often house a designated clinic that provides neuropsychologists with the opportunity to supervise students as they begin providing clinical services to patients. This role allows us to continue to perform some clinical work while primarily focusing on teaching and research. Alternatively, faculty members who are interested in climbing the professional ladder can work their way up into administrative positions (e.g., Department Chair, Dean of Arts and Sciences). This is similar to the business sector, where someone might start out performing a specific role (e.g., lead baker in a deli) but may end up in a general executive role (e.g., store manager or district manager). The advantages to these promotions include greater influence, financial compensation, and prestige. The disadvantages can include additional politics and stress, as well as less time available for one's original passions (e.g., teaching and/or research).

Hospitals

What images come to mind when you envision a hospital? Do you imagine a bustling, chaotic intensive care unit (ICU), with doctors and nurses frantically hustling from patient to patient, treating serious injuries and ailments, and providing lifesaving care? Or do you think about a quiet, tense operating room (OR), where a dozen surgeons and hospital staff are working in choreographed synchrony, carefully removing seizure-generating brain tissue? Or maybe we are way off here, and your idea of a hospital is a slower-paced outpatient clinic, with patients sitting in the waiting room, reading magazines, and checking their text messages, waiting to be seen by the doctor who can diagnose and treat their ailments. If any of these scenes come to mind when you think about a hospital, the images could have included a neuropsychologist. You might find us completing mental status examinations for people with severe traumatic brain injuries (TBIs) in the ICU, performing a Wada test in preparation for epilepsy surgery,[3] or testing and treating patients with a plethora of different brain diseases in outpatient clinics.

There are various types of hospitals, and we will cover those that are most relevant in this section. First, however, we will outline the basics of medical neuropsychology. In hospitals, neuropsychologists are often viewed by patients as similar to medical doctors such as radiologists and cardiologists, and although we are not physicians, our training in both neurobiology and psychology allows us to effectively straddle the worlds of medicine and psychology. In these settings, we are specialty healthcare providers, and we receive referrals from physicians in order to answer important clinical questions. These referrals vary greatly depending on the patient population and can include characterizing a patient's cognitive strengths and weaknesses, providing recommendations for discharge (e.g., to a rehabilitation facility or a home environment), and engaging in treatment planning (e.g., psychotherapy and/or cognitive training). We also see a large diversity of different patient populations at medical centers. We perform neuropsychological evaluations and interventions for people with Parkinson's disease, Huntington's disease, brain tumors, epilepsy, multiple sclerosis, multiple system atrophy, gunshot wounds, organ failure, organ transplants, endocrine dysfunction, and many others. The term we used earlier – medical neuropsychology – essentially refers to the impact of any biomedical process on our cognitive abilities. Some of these processes may seem obvious (e.g., stroke), but others may not (e.g., kidney failure). Clearly, neuropsychologists are unique, and we have important roles in hospital settings.

****

At the beginning of this section, we provided a few examples of activities that neuropsychologists perform at hospitals. These activities can be broadly organized into two categories: *inpatient* (the patient is currently residing at the hospital) and *outpatient* (the patient is coming in for an appointment and then leaving). Inpatient neuropsychological assessments occur in the acute (immediately following) or post-acute phase of a brain insult. In this role, we are mobile assessors, traveling around the hospital from room to room, often testing patients while they sit in their hospital beds. Our inpatient evaluations tend to be shorter in length, the environment tends to be chaotic (e.g., tired, distracted patients, interruptions from nurses and family members), and our testing is often designed to measure the patient's current cognitive functioning in order to engage in treatment and discharge plan-

---

[3] A Wada test is a semi-invasive procedure in which half of the brain is anesthetized for a short period of time in order to assess the functioning of the other hemisphere in isolation.

ning. For example, imagine that a 10-year-old girl fell off of her bicycle and sustained a TBI four days ago. Our evaluation can provide information regarding the extent of cognitive deficits caused by her injury and can help her physicians decide on the best course of action. We might help answer multiple clinical questions, including, "Would she benefit from occupational therapy to help her recover her functional and academic skills?" and "Are her cognitive difficulties mild enough that she can be discharged home to begin attending school again in the near future?"

The girl in the above example may subsequently be followed by a neuropsychologist as an outpatient. Here, patients are residing in their homes, and they visit our neuropsychological clinic (housed within the hospital) in order to be assessed and/or treated. Advantages of outpatient compared to inpatient testing include more stable patients who provide more reliable test data (because they are not acutely sick or injured), control over the testing environment, and additional time to go into greater detail in our assessment batteries. The majority of neuropsychological evaluations are conducted in outpatient settings, and so these are our bread-and-butter assessments.

* * * *

In addition to different settings within hospitals, there are many different types of facilities. The following are some of the most common categories under the "hospital" umbrella.

Academic Medical Centers

As the name implies, academic medical centers (AMCs; also occasionally referred to as "university-affiliated hospitals") are important hubs for medical research, and many are among the most well respected in the US. For a few examples, Massachusetts General Hospital is affiliated with Harvard Medical School, Barnes Jewish Hospital is affiliated with Washington University in Saint Louis, Oregon Health and Science University (OHSU) Hospital is affiliated with OHSU, and Ronald Reagan UCLA Medical Center is affiliated with UCLA. AMCs typically offer all of the same clinical services as a "freestanding hospital" (a medical center without a university affiliation), plus they boast large research programs as well. Many practicing physicians and neuropsychologists at AMCs were trained by the associated university, and the services they provide tend to be at the cutting edge of scientific progress and innovation.

The research conducted at AMCs differs from that which is completed in a non-affiliated university setting in that it is typically more clinically relevant. In other words, the research questions that inform these projects are asked in order to better understand a patient population and/or to discover treatments with a direct impact on patients. This is sometimes called "bench to bedside" – the idea that we take findings from a scientific laboratory (the bench) and use them effectively in real-world clinical settings (the bedside). Moreover, the flip side is also true: the clinical work conducted in AMCs often feeds back to inform the research questions that are asked. For example, working with stroke patients at the hospital gave me (John) the idea for my dissertation project, where I tested cognitive recovery following pomegranate supplementation in post-acute stroke patients. I assembled a team of neuropsychologists, neuroscientists, physicians, physical therapists, and occupational therapists in order to approach the trial from different angles. We found that patients who received the supplements during their hospital stay had better cognitive and functional recovery and were discharged from the hospital quicker than were patients who received a placebo (Bellone et al., 2018). It is important to note that this was only a pilot study (i.e., we used a very small sample of patients to determine whether a larger trial would be worthwhile) and that the findings need to be replicated and expanded before this intervention is widely adopted, but I still recommend pomegranate products whenever I get the chance!

As we mentioned earlier, training is a crucial aspect of AMCs. Local medical schools and psychology programs use AMCs to train the next generation of physicians and psychologists through practicum placements, internships, and residencies. Neuropsychologists housed in AMCs typically have a new batch of students to supervise every 1–2 years. These students can bring fresh ideas, energy, and excitement that strengthen the service. They shadow neuropsychologists as they learn more about our assessment tools and procedures. Across time, they become more skilled in our trade, they gain more independence, and they take on additional responsibilities. One model for this is the "see, then do, then teach" method of learning, and AMCs offer all three. That is, trainees can observe a neuropsychologist at their job, perform that role themselves, and then teach other students. Overall, AMCs are ideal training sites for several reasons: (a) they often have cutting-edge tools and methods, (b) they attract patients with the most difficult-to-treat/rare conditions, and (c) they offer an abundance of research opportunities.

Related to training, an additional by-product of the proximity between hospital and university is a highly intellectually stimulating environment. AMCs are often teeming with didactics, seminars, and informal learning opportunities available to faculty, staff, and students. For example, neuropsychologists and trainees sometimes have the opportunity to attend "brain cuttings," where they observe a neuropathologist perform an autopsy in order to help determine the cause of death and the extent of brain disease. The very next day they might observe a neurosurgeon resect a brain tumor, or they might attend neuroradiology rounds in order to participate in vibrant discussions about results of various imaging techniques such as MRI, positron emission tomography (PET), and computed tomography (CT). These opportunities often seem to fall from the sky in AMCs, and it does not take long before you will be forced to begin turning down interesting training opportunities because, unfortunately, there are only 24 hours in a day!

Veterans Affairs (VA) Medical Centers

What is your impression of the VA healthcare system in the United States? Chances are, if you watch the news, you may think that the VA system is broken, that the doctors and staff who work there are all incompetent, and that our nation's Veterans are mostly left to fend for themselves. Fortunately, this is not the case. Certainly, a number of VA hospitals have significant problems, and we always want to improve the care we are providing to our Veterans. But, as in other cases, extreme ideas put forth by the media can be misrepresentations of reality. All VA hospitals that we have encountered are staffed by well-trained, highly skilled professionals. They boast cutting-edge research and technology, they train the next generation of healthcare providers (including neuropsychologists), and they provide consistent, high-quality care to those who have courageously served our country. As an example of some of the innovation taking place in this system, the VA led the way in the early phases of the "telehealth" movement. Telehealth refers to the delivery of healthcare services remotely, to people who are unable to travel to the nearest major medical center, which could be hundreds of miles away. Telehealth capitalizes on videoconferencing software such as Zoom to allow for remote communication between patient and provider, and it tends to benefit those who are in the greatest need, such as people of low socioeconomic status and people living in rural communities.

In our field, neuropsychologists are adopting telehealth and labeling it "teleneuropsychology" (Cullum et al. 2014).[4] In fact, we (John and Ryan) have both conducted full teleneuropsychological evaluations with Veterans. Our patients tend to find teleneuropsychology to be quite useful; they often think about it as Skyping or Facetiming with their doctor, and it is much more convenient and practical for them (Parikh et al. 2013). We also love it because we are able to provide services to people who would otherwise go without help. Overall, we are proud that the VA system is spearheading this frontier in healthcare service delivery, and it is just one example of the important initiatives designed to improve the quality of life for our nation's Veterans.

Importantly for our purposes, VA medical centers are typically academic consortiums (i.e., each VA is affiliated with a university), and they function similarly to AMCs with the obvious exception that they specialize in treating Veterans. Broadly speaking, the VA operates as a federal system, so you might assume that all VA hospitals are essentially the same. However, you would be wrong. An important rule of thumb when considering a particular VA environment is that "if you've seen one VA... you've seen one VA." In other words, despite being bound together by federal rules and regulations, each of these hospitals (170 at last count) is its own unique entity with its own culture, methods, and personality. That being said, there are a few characteristics of all VAs that are important to be aware of. For example, mili-

---

[4]Teleneuropsychology is becoming even more relevant and useful in the wake of the COVID-19 crisis, where physical distancing has led to the widespread adoption of remote cognitive testing via teleconferencing platforms (www.NavNeuro.com/41).

tary culture is pervasive across sites, and it behooves neuropsychologists working in this setting to learn basic military lingo, including ranks, branches, roles, etc. Understandably, Veterans are typically very proud of their branch of the military (often wearing it, literally, on their sleeve) and of the fact that they served their country. Finally, Veterans often feel more comfortable participating in activities and support groups with other Veterans because they have a shared experience and identity. Consequently, VAs are more than simply medical centers providing care to Veterans. They act as spaces for military tradition and culture, as well as refuges of respect, comfort, and support for these men and women.

****

The demographic characteristics of patients at VA hospitals are different in comparison to other medical centers in several ways. For example, there is a bimodal age distribution, with most individuals either being older adults from the Vietnam and Korean War eras or younger adults from the conflicts in Iraq and Afghanistan. The population is primarily male, mostly White, and typically more medically complex than is encountered in other settings. In other words, many Veterans have systemic health problems with wide-ranging impacts on their overall well-being; these conditions include obesity, diabetes, hypertension (high blood pressure), hypercholesterolemia (high cholesterol), depression, and anxiety, among others (Agha et al. 2000; Das et al. 2005). The high prevalence rates of these issues are very relevant for neuropsychologists because they all have the potential to negatively impact brain function, potentially leading to cognitive impairments (Debette et al. 2011; Yaffe et al. 2010).

To contextualize these ideas, imagine that you were drafted to fight in the Vietnam War. You were taken from your home and family, shipped 8000 miles away, and placed in the center of terrifying and horrific events. You sustained physical injuries (e.g., shrapnel from a nearby explosion). You were exposed to opiates, cocaine, and psychedelics. And then you came home to a society that largely vilified you for having served in the military. On top of that, you began experiencing mood changes, nightmares, flashbacks, and intense emotions associated with the events you witnessed overseas (symptoms of posttraumatic stress disorder [PTSD]), and then you started consuming alcohol to attempt to dull the pain. The prolonged psychological stress, alcohol use, insomnia, and physical discomfort takes a major toll on your body and brain, and you eventually develop chronic medical, psychiatric, and neurocognitive conditions. You visit a neuropsychologist in order to try to understand why your attention, memory, and problem-solving abilities have declined significantly. From this example, we hope that we have made it clear why it is so common for Veterans to have a broad range of ailments and why they visit neuropsychologists. Luckily, our country and the medical community at large have come a long way in terms of caring about and caring for Veterans since the time of the Vietnam War.

On the other side of the age spectrum, young adults returning home from the arid deserts of the Middle East had a vastly different experience than did those who fought in the dense tropical vegetation seen in the jungles of Vietnam. On October 7, 2001, President George W. Bush announced the commencement of airstrikes against the terrorist groups responsible for the September 11 attacks on the World Trade Center (the Taliban and Al Qaeda). This conflict took place primarily in Afghanistan and has been labeled Operation Enduring Freedom (OEF). Two years later, in what is referred to as Operation Iraqi Freedom (OIF), the United States and its allies invaded Iraq and ultimately toppled the government of Saddam Hussein. These armed conflicts have been controversial for many reasons, including the loss of life and injuries suffered by American soldiers and innocent civilians. For example, according to the US Department of Defense (https://dod.defense.gov/News/Casualty-Status/), there were 4410 military casualties in OIF, and 31,957 soldiers were wounded in action. Of those people who were wounded in OIF/OEF, many sustained TBIs from improvised explosive devices (IEDs, a type of bomb). That is, while many civilian TBIs occur from a mechanical impact such as a hard hit in a football game or the force of one's head striking the steering wheel in an automobile accident, many Veterans of OIF/ OEF sustained brain injuries from the explosions and subsequent force of an IED (called "blast injuries"). Neuropsychologists evaluate Veterans who are experiencing ongoing cognitive difficulties that may be partially attributable to these TBIs. The Veterans also frequently have symptoms of PTSD, sometimes from the same event that caused the TBI. We pride ourselves in not only performing cognitive testing to elucidate areas of weakness but also in providing psychoeducation and cognitive training interventions to help these Veterans improve their attention, memory, and problem-solving abilities so that they can better cope with their injuries and improve their quality of life.[5]

Rehabilitation Hospitals

Rehabilitation hospitals are medical facilities that specialize in helping people recover after an acquired injury or an acute medical condition significantly impacts their ability to care for themselves. Many facilities have special units or wards that serve a specific population such as a stroke unit, a spinal cord injury unit, or a TBI unit. Once the individual is medically stabilized, they are transferred to one of these units to receive daily medical supervision and a suite of therapies (e.g., physical therapy, speech therapy, occupational therapy, therapeutic recreation). This is typically done on an inpatient basis, with hospital stays ranging from days to weeks depending on patient and injury characteristics (e.g., health insurance, space at the facility, the patient's needs and wishes, the discharge plan).

Rehab units frequently employ interdisciplinary teams that meet regularly to discuss each patient's needs and coordinate care. While these teams exist in a wide

---

[5]If you are interested in learning more about working with Veterans, check out *Psychological assessment of Veterans* (2014) by Shane Bush.

variety of medical settings, they are especially prevalent in rehab hospitals. The team may consist of an attending physician, resident physicians, therapists from the aforementioned disciplines, a social worker, nursing staff, and a neuropsychologist and/or general psychologist. Interdisciplinary teams are outstanding models for patient care. You can think of these teams as analogous to basketball teams, where each player has a specific skillset and role. On a great basketball team, the individuals complement one another, compensating for limitations and weaknesses of individual players and operating in synchrony as a single effective unit. The whole is far greater than the sum of the individual parts. This is what we strive for as interdisciplinary teams. Each of us brings our expertise to the table and helps the team understand the patient from a particular vantage point. For example, the attending physician (often a physiatrist) understands and elucidates the biological underpinnings of the patient's injury, as well as the medical procedures and interventions that can improve physical ailments. The social worker is the expert in the patient's broader environment and context such as their occupational, family, and housing status. The neuropsychologist determines the patient's cognitive and psychological status, tracks cognitive recovery over the course of the hospital stay, helps others understand how best to communicate and interact with the person (e.g., "Can she comprehend and remember verbal instructions?"), plans for discharge needs (e.g., "Will he be capable of managing his own medications?"), and provides other recommendations to compensate for potential cognitive deficits.

Neuropsychologists in this setting frequently conduct cognitive assessments, but we also engage in much more intervention work than is common elsewhere. We often provide cognitive training to target specific areas of impairment. We perform ongoing individual psychotherapy to help people cope with the low mood and high degrees of stress and anxiety that often accompany these injuries. And family therapy is also typically helpful because these injuries affect more than simply the patient – they impact the entire family system. For example, imagine that a 40-year-old man who has been the patriarch and breadwinner of the family sustains a massive stroke and subsequently requires full-time assistance from his wife and children. Such an event would result in a drastic role rearrangement, as well as alterations to each person's self-identity. The wife might transition from partner to caregiver. The eldest daughter may feel obligated to find a part-time job and start babysitting for her younger sibling. The patient himself may struggle with his inability to work, ongoing physical limitations, trouble communicating, and general loss of independence. All of these issues are very difficult for people to cope with, and the clinicians at rehab hospitals are on the front lines in terms of preparing patients and helping them cope with such significant life changes.

Another type of injury that leads to a large shift in identity and family dynamics is a spinal cord injury. Neuropsychologists serving on a spinal cord injury team focus their work in assessing and treating the psychological difficulties that occur in this population and neuropsychological sequelae of comorbid TBIs. As you can imagine, an injury to the spinal cord that leads to the loss of sensation and movement of your lower limbs (paraplegia) or your upper and lower limbs (quadriplegia) is a devastating event with which to cope. For people who are currently able-bodied, it is impossible to fully grasp what this experience must be like. Given the gravity of these injuries, we are not surprised that people struggle with depression, anxiety, and emotional stress. Neuropsychologists strive to help patients with spinal cord injuries learn to adjust to their new life as well as they possibly can.

It is not an accident that neuropsychologists spend so much of their time at rehab hospitals providing treatment and intervention. A big factor driving this phenomenon is the nature of the injuries seen in this setting. Stroke and TBI, both common in rehab settings, differ from neurodegenerative conditions (e.g., Alzheimer's disease) in that the injury is typically a single, discrete event, rather than a progressive disease process. This allows for recovery following the initial brain insult. Indeed, there exists a "therapeutic window" where the potential for recovery is greatest, and therefore, effective intervention is paramount (Cramer 2008).

To take a TBI case as an example of the recovery process, during my practicum, I (John) evaluated a young man who suffered a moderate TBI following a pole vaulting accident – he missed the mat and came down head first on the hard turf. He was admitted to our TBI unit about 1 week after the accident (he first needed to be stabilized in the intensive care unit). I visited him many times over the course of his 3-week rehab stay, watching him progress from very confused and agitated to the calm, personable young man his family knew him to be. He initially did not have any sense about what year it was, where he was, or why he was at the hospital, but he gradually became more and more oriented[6] as the days went on. Testing him serially throughout his stay, I watched his cognitive abilities evolve from severely impaired to mildly impaired, and I helped him and his family prepare for how to deal with the cognitive challenges when he returned home. I also prepared them for what to expect; that is, that he would likely make further cognitive and functional gains over the course of the next several months but that it was uncertain whether he would recover back to his baseline. I always take a "hope for the best, prepare for the worst" perspective and encourage patients and families to do the same. Armed with numerous recommendations, community resources, and a solid discharge plan, it was an honor to see him walk out of the hospital in a much better state than when he was wheeled into it, knowing that I had a small part to play in this successful outcome.

Psychiatric Hospitals

In the popular 1975 movie, *One Flew Over the Cuckoo's Nest*, Jack Nicholson plays the arrogant, cynical Randle McMurphy who is placed in a psychiatric hospital despite the fact that he is not mentally ill (he acted the part in order to avoid jail time). Upon admission, he finds himself in a horrific environment where the residents are treated more like prison inmates than hospital patients in need of help. His

---

[6] By oriented, we mean aware of the where, when, and why information.

misbehavior and constant run-ins with the cold, heartless nurse Ratched lead to increasingly punitive measures. Ultimately, after he ropes other patients in on some of his truancy and general mischief, he is "sentenced" to electric shock therapy, where he is strapped down to a table and electrocuted repeatedly as he slowly turns into a zombie-like creature.[7] The movie left many people with chills when they considered the prospect of admitting themselves or their loved ones into one of these institutions, and *One Flew Over the Cuckoo's Nest* is far from the only depiction of psychiatric hospitals as hellish environments. Indeed, there are many other books, movies, and TV shows that portray the conditions of psychiatric hospitals, and the idea of these places as brutal, torturous environments has slowly worked its way into our zeitgeist.

It is not surprising that Hollywood latched onto the notion of psychiatric hospitals as frightening places. In the nineteenth and early twentieth centuries, people who were severely mentally ill (in particular, people with schizophrenia) were labeled as "mad" or "lunatics" and were often thrown into so-called "insane asylums" in order to shield the rest of society from them. There are accounts of these institutions that sound horrific, even worse than what was depicted in *One Flew Over the Cuckoo's Nest*. In fact, these institutions were so overcrowded that extreme measures were taken in an effort to discover and disseminate a "cure" for severe mental illness. For example, transorbital lobotomies (inserting a surgical tool up through the eye socket in order to destroy portions of the prefrontal cortex) were routinely performed in order to reduce agitation and paranoia. However, the procedure caused immense harm to thousands of people who were left permanently brain damaged and unable to think and function normally. Fortunately, with the advent of antipsychotic medications in the 1950s, people with psychotic disorders began to receive actual treatments, and many were ultimately deinstitutionalized and released back into society as their symptoms were partially alleviated. For those who were still symptomatic and remained hospitalized, the focus slowly shifted to ensuring the best quality of life rather than merely keeping them locked away.

Luckily for us, most psychiatric hospitals in today's world do not resemble the mental institutions of the past. Modern versions are designed to assess and treat people with significant mental health problems, with the ultimate goal of helping them recover and achieve a high level of functioning in society. There are a number of characteristics of these facilities that are quite relevant. First, as you might imagine, patients receiving care in psychiatric hospitals may present with a wide range of emotional and psychological disorders, including depression, bipolar disorder, schizophrenia, generalized anxiety disorder, panic disorder, agoraphobia, post-traumatic stress disorder, obsessive compulsive disorder, anorexia nervosa, and many others. Second, psychiatric hospitals often have an inpatient wing for people who are acutely ill (e.g., someone with severe depression who is actively suicidal),

---

[7] "Shock therapy," more appropriately termed electroconvulsive therapy (ECT), is still used in psychiatric hospitals but in a much safer manner than in the past. It is not a first-line treatment because there can be transient memory loss, but it can be highly effective for severe mood and psychotic symptoms (e.g., Kellner et al. 2012).

as well as outpatient and intensive outpatient services. Third, they are typically structured based on the age cohort of the patient (i.e., a child/adolescent unit, an adult unit, and a geriatric unit).

Neuropsychologists are right at home in this setting because much of our early training is in psychopathology, assessment, and treatment. Most people are unaware that any and all conditions that impact a person's mental health also have a great potential to impact thinking abilities. That is, someone who is struggling with low mood, anxiety, stress, compulsive behavior, or psychosis (e.g., hallucinations or delusions) is usually also struggling with deficits in attention, processing speed, memory, and executive functions. One reason for this is that when our bandwidth is taken up by one process (e.g., overwhelming anxiety), we have fewer cognitive resources to use for other tasks (e.g., remembering our work schedule). Consequently, patients in psychiatric hospitals often benefit from neuropsychological evaluations to identify their cognitive impairments and to provide treatment recommendations.

* * * *

In comparison to other settings, the role and identity of a neuropsychologist in a psychiatric hospital is quite different. For example, in a neurology department housed in an AMC, the neuropsychologist is frequently interacting with other experts in traditional brain disorders. The other doctors have extensive knowledge and skills in neuroimaging, neuropathology, and the biology of many neurological conditions. Neuropsychologists are sought after for our expertise in (a) assessment of cognition and (b) knowledge of mental illnesses. In contrast, in a psychiatric hospital, we work with psychiatrists and other healthcare professionals with years of training and experience treating mental disorders. Here, we are not the go-to experts in mental illness because others are also well-qualified in this area; rather, we are often consulted for our expertise in cognitive assessment and in the brain itself. This state of affairs is not necessarily an advantage or a disadvantage – it is simply a difference. It speaks to the idea discussed in Ch. 2 that the breadth of our training allows us to wear many different hats and to take on a variety of roles in different settings.

Private Practices

Although private practice is not what most people think of when they imagine where they might find a neuropsychologist, it is actually the setting in which about half of neuropsychologists find themselves (Sweet et al. 2020). Typically, "private practice" or "group practice" means that a neuropsychologist is not part of a larger entity (e.g., a university or hospital) but instead functions independently or in a group of other providers. The most traditional venue is an office suite (think of a dentist's office…without the drills, chisels, and laughing gas), although the practice can also be housed within a hospital, in the practitioner's home,[8] or it could even be mobile.[9]

Regardless of the brick and mortar location, a private practice is a business, so a neuropsychologist starting up his or her own practice is also an entrepreneur. This has pros and cons, as you might imagine. One of the greatest benefits is flexibility. If you own a private practice, then you are your own boss, with virtually unlimited autonomy and freedom, and you don't have pesky hospital regulations, trainings, and oversight to hold you back. As such, you can create your own work schedule, set up your environment to your preferences, determine the types of services to provide and the assessment measures to use, format neuropsychological reports as you see fit, and otherwise conduct business your way. The flexibility regarding the number of patients you schedule, hours you work, and types of services you provide means that there is a higher ceiling in terms of how much money you can make. The more patients you see, the more you earn. Furthermore, the location is flexible. Because you can open up a practice anywhere, you are not bound by the location of the nearest large medical center, and you have the option of living outside of a city, which often means cheaper cost of living and reduced daily commute time.

However, there is a flip side to the coin, and working in a private practice comes with disadvantages. As the clinician, you are typically not salaried, so if you choose to accept fewer patients because you want to leave work early and have a long week-end, then you do not earn money during that time. Your income also depends on the referrals, which can ebb and flow dramatically so that it is difficult to know exactly how much money to anticipate each pay period. Similarly, income often depends on whether insurance reimburses you for the services you provided, and negotiating with insurance companies can be quite time-consuming and frustrating. Furthermore, the benefits (e.g., paid time off, health insurance, retirement plans) tend to be less generous than would be found in a large institution, if present at all. Finally, the fact that you are not working for a university-affiliated hospital means that you are missing out on the academic and intellectual opportunities offered by these organizations.

As the previous paragraphs have made clear, with flexibility and autonomy comes responsibility. As the owner of a private practice, you have all of the typical duties of a neuropsychologist, together with the burden of running a small business. This includes managing finances and accounting, infrastructure (e.g., finding a work space, setting up computer stations and networks, storing medical records), marketing your services, and administration (e.g., calling/scheduling patients, submitting insurance claims), among others.[10] Going this route can be very challenging, especially at first, given the many details that need to be ironed out and the associated start-up costs. In order to limit expenses and complexity, many practitioners start

---

[8] There are ethical, safety, and privacy concerns to consider when conducting in-person evaluations or treatment out of the home.

[9] Mobile assessment could involve a neuropsychologist traveling around to assess older adults in skilled nursing facilities.

[10] For more details, read *The Business of Neuropsychology* (2010) by Mark Barisa.

out small and slowly build up the practice by increasing the client load, buying more test materials, and expanding or improving the environment. They might hire office staff (e.g., secretary, billing specialist) and other clinicians as the practice grows so that they can off-load some of those administrative duties and service more patients. Some groups grow large enough that they almost feel like small hospitals in and of themselves and may offer many of the luxuries (e.g., health insurance, trainings/ seminars) but also some of the downsides (less freedom) of other settings. In this scenario, the founding owner would likely see fewer and fewer patients of their own and would move into more of a managerial position, as "CEO" of the practice.

One way to mitigate many of the aforementioned disadvantages of a private practice is to join an existing group rather than starting your own from the ground up. This avoids the responsibility of owning the company, although a percentage of the fees you collect will go to the practice itself. Still another option is to co-create a practice with other neuropsychologists or a mix of other professionals (e.g., psychiatrist, general psychologist) so that the tasks and expenses are distributed and the different specialists can refer patients to one another when appropriate.

****

Before concluding this section, we want to mention that there is a specific type of neuropsychological assessment that is frequently associated with private practice: forensic work. This involves conducting evaluations when there is a legal component, as opposed to performing an assessment strictly to improve the health and welfare of the patient. There are many different types of cases under this "forensic" umbrella, but they generally fall under two categories: criminal and civil. For example, in the criminal arena, an attorney could bring in a neuropsychologist in order to provide an opinion as to whether a defendant is cognitively capable of standing trial for his alleged crime. On the other hand, one type of evaluation that falls under the civil umbrella is referred to as "personal injury," where someone sues another person after a car accident or other event in which an alleged neuropsychological disorder was sustained, and it is the job of the neuropsychologist to determine whether cognitive or psychiatric issues are related to the accident. Additionally, in workers' compensation cases (civil), an employee attempts to procure wages and/or medical expense reimbursement following certain work-related injuries. A neuropsychologist may be brought in by either side (employer or employee) to determine the existence or nonexistence of cognitive or neurobehavioral deficits as a result of the injury.

Until now, we have been discussing clinical work in a variety of different environments (VA hospitals, psychiatric hospitals, etc.), but forensic work is qualitatively different from clinical work and has its own unique pros and cons. Specifically, advantages include higher compensation rates and the opportunity to become involved in the analysis of novel cases, which are often complex due to the legal component. Downsides include challenges of navigating the legal system, which by its nature is adversarial, and the fact that we are not able to put all of our energy into helping the client, as this is not our job in this role. Fortunately, we do not have to choose either clinical work on the one hand or forensic work on the other. Many neuropsychologists pick up a small number of legal cases while still retaining their day jobs.[11] In this context, we think that forensic work can be a fun and rewarding adjunct to our more traditional roles.[12]

Conclusion

We have learned throughout this chapter that there are advantages and disadvantages to every work setting in neuropsychology.[13] What is especially important is that you find the environment that best suits your needs and desires. One excellent strategy is to seek out multiple different settings during your training so that you have a sampling of what it would feel like to work in a variety of arenas. This way, when it comes time to pursue a career, you will have a plethora of experiences under your belt, and you will know the pros and cons to each as they apply to you. That part is key – although we tried to hit the high points of each of these settings in order to arm you with important information, each and every one of us has our own idiosyncratic preferences, strengths, and weaknesses, and there is no substitute for personal experiences. So, if you are interested in neuropsychology, diversify your training and be sure to pause and soak everything in as you move through your training. We hope that you have as much fun with it as we did!

> You really have to enjoy the ride. Get jazzed about this stuff! Enjoy the process, not just the outcome.
>
> – Robert D. Latzman, PhD

References

Agha, Z., Lofgren, R. P., VanRuiswyk, J. V., & Layde, P. M. (2000). Are patients at Veterans Affairs medical centers sicker?: A comparative analysis of health status and medical resource use. *Archives of Internal Medicine, 160*(21), 3252–3257.

Bellone, J. A., Murray, J. R., Jorge, P., Fogel, T. G., Kim, M., Wallace, D. R., & Hartman, R. E. (2018). Pomegranate supplementation improves cognitive and functional recovery following ischemic stroke: A randomized trial. *Nutritional Neuroscience, 22*(10), 738–743.

Bigler, E. D. (2019). Neuroimaging and neuropsychology. In K. M. Sanders (Ed.), *Physician's field guide to neuropsychology* (pp. 421–434). Springer.

---

[11] However, there is extra training/education that is needed to be proficient in forensic settings.

[12] For more information on this topic, we recommend the book, *Forensic Neuropsychology* by Glenn Larrabee.

[13] We did not have room to describe *every possible* setting in which neuropsychologists sometimes work, so we highlighted the most common settings. Other settings include industry (i.e., private corporations such as pharmaceutical companies or test developers), sports medicine clinics and athletic teams (e.g., John is a neuropsychology consultant for the Anaheim Ducks NHL team), the military, correctional facilities, and schools.

Bush, S. (Ed.). (2014). *Psychological assessment of Veterans*. Oxford University Press.

Cramer, S. C. (2008). Repairing the human brain after stroke: I. Mechanisms of spontaneous recovery. *Annals of Neurology, 63*(3), 272–287.

Cullum, C. M., Hynan, L., Grosch, M., Parikh, M., & Weiner, M. (2014). Teleneuropsychology: Evidence for video teleconference-based neuropsychological assessment. *Journal of the International Neuropsychological Society, 20*(10), 1028–1033.

Das, S. R., Kinsinger, L. S., Yancy, W. S., Jr., Wang, A., Ciesco, E., Burdick, M., & Yevich, S. J. (2005). Obesity prevalence among veterans at Veterans Affairs medical facilities. *American Journal of Preventive Medicine, 28*(3), 291–294.

Debette, S., Seshadri, S., Beiser, A., Au, R., Himali, J., Palumbo, C., et al. (2011). Midlife vascular risk factor exposure accelerates structural brain aging and cognitive decline. *Neurology, 77*(5), 461–468.

Gajda, R., & Tulikangas, R. (2005). *Getting the grant: How educators can write winning proposals and manage successful projects*. Association for Supervision and Curriculum Development.

Kellner, C. H., Greenberg, R. M., Murrough, J. W., Bryson, E. O., Briggs, M. C., & Pasculli, R. M. (2012). ECT in treatment-resistant depression. *American Journal of Psychiatry, 169*(12), 1238–1244.

Larrabee, G. J. (2012). *Forensic neuropsychology: A scientific approach* (2nd ed.). Oxford University Press.

Parikh, M., Grosch, M. C., Graham, L. L., Hynan, L. S., Weiner, M., Shore, J. H., & Cullum, C. M. (2013). Consumer acceptability of brief videoconference-based neuropsychological assessment in older individuals with and without cognitive impairment. *The Clinical Neuropsychologist, 27*(5), 808–817.

Robertson, J. D., Russell, S. W., & Morrison, D. C. (2020). The grant application writer's workbook: National Institute of Health version. Grant Central, LLC.

Sweet, J. J., Klipfel, K. M., Nelson, N. W., & Moberg, P. J. (2020). Professional practices, beliefs, and incomes of US neuropsychologists: The AACN, NAN, SCN 2020 practice and "salary survey". *The Clinical Neuropsychologist*, 1–74.

Yaffe, K., Vittinghoff, E., Lindquist, K., Barnes, D., Covinsky, K. E., Neylan, T., et al. (2010). Posttraumatic stress disorder and risk of dementia among US veterans. *Archives of General Psychiatry, 67*(6), 608–613.

Zlowodzki, M., Jönsson, A., Kregor, P. J., & Bhandari, M. (2007). How to write a grant proposal. *Indian Journal of Orthopaedics, 41*(1), 23–26.

Chapter 4

Challenges to Working in Neuropsychology

As with any profession, there are inherent challenges we must face, both on the road to becoming neuropsychologists, and when engaging in this work on a daily basis. However, I can't help but notice that some of the most salient challenges offer the most rewarding experiences.

– Meghan Collier, PhD

We have presented a pretty rosy picture of neuropsychology thus far, mainly because we both feel strongly that there are many positive aspects to working in this field and we want you to be aware of the upsides. However, every occupation has its downsides, and we would not be doing our jobs if we gave you the impression that neuropsychology is rainbows and sunshine every second of every day. Before you fully commit to this path, you should be aware of the aspects of the job that are less than pure delight.

Challenges present themselves both during training and in daily life as a professional. We will attempt to provide you with guidance regarding many specific training hurdles as we encounter them in Part II. We also just described issues specific to particular professional settings in Ch. 3. Here, we list some of the overarching challenges that are common either during our training, after we settle into a professional position, or both. We offer advice where we have it, but some of these are simply unavoidable by-products of the job description. We will do our best to be honest and balanced, in order to help you make the best possible decision for your future. With that in mind, we cover the following topics in this chapter: (1) *the long, winding road*, (2) *wearing multiple hats*, (3) *ambiguity*, (4) *academic challenges*, and (5) *clinical challenges*.

© Springer Nature Switzerland AG 2021
J. A. Bellone, R. Van Patten, *Becoming a Neuropsychologist*,
https://doi.org/10.1007/978-3-030-63174-1_4

Table 4.1 Typical number of years spent at each stage for the different trajectories

	College	Graduate or medical school	Residency	Fellowship	Total after high school
Medicine	4	4	3–7	0–2	11–17
Neuropsychology	4	4–5	1	2	11–12

The Long, Winding Road

Most people are more familiar with the training required to become a physician than they are with the training required to become a neuropsychologist. While the content of the training differs between these two fields, the steps required are broadly similar (see Table 4.1). Medical doctors typically complete a 4-year college degree, 4 years of medical school, 3–7 years of residency (depending on their specialty), and an optional fellowship. As noted in the Preface, neuropsychologists also graduate from college and then complete about 5 years of graduate school, 1 year of residency (more commonly called "internship"), and 2 years of postdoctoral fellowship. In both cases, there are many years of study, at least a bit of sleep deprivation, and thousands of hours spent learning and building expertise.

From a purely financial perspective, there is clearly an opportunity cost to this time spent on your education. That is, if you are training for this long, you will be missing out on years of potential earning power, when you could be working a job and making a competitive salary rather than accruing more student loans. When I (John) told my dad that I had decided to pursue a career in neuropsychology and laid out the timeline and cost of school/training, he blurted out, "If you're going to spend all that time and money in school, you might as well go into medicine!" He believed that the only way to justify that many years of training was to ultimately earn hundreds of thousands of dollars per year as a physician. Fortunately, he eventually came around when I explained that maximizing earning power was not my primary motivation and I hit him with all of the "Why Neuropsychology?" reasons we laid out in Ch. 2.

One additional potential opportunity cost of the long, windy road pertains to family life. Although many people successfully begin a family during their training, many others decide to put it on hold until they are hired at their first job, for the reasons discussed in this section. Ultimately, this decision is made by each person based on their own personal circumstances.[1]

****

---

[1] In 2018, 13% of psychology internship applicants said they had dependent children living with them (https://www.appic.org/Internships/Match/Match-Statistics/Applicant-Survey-2018-Part-1). Although this can pose quite a challenge, many people have successfully navigated graduate school while simultaneously raising children.

Related to the time and effort spent earning your education, training in neuropsychology is often a windy road from a geographic perspective. Let's take an example to prove the point. I (Ryan) completed my undergraduate degree in Harrisonburg, VA. I then moved to Saint Louis, MO, for 5 years of graduate school, Providence, RI, for 1 year of residency, San Diego, CA, for 2 years of postdoctoral fellowship, and then to Boston, MA, for my first grown-up job. Without a doubt, I have become well-acquainted with cardboard boxes, moving trucks, furniture pads, and mail forwarding. Many other people have similar stories. And while it is sometimes possible to stay in the same location for two steps in a row, most of us have moved at least two or three times from start to finish.[2]

Of course, moving can be stressful and expensive, and it's not most people's idea of an exciting weekend. On the other hand, there are nice benefits such as the opportunity to live in different geographic regions, to network with a variety of colleagues and supervisors, and to explore everything that new communities have to offer. It is not often in life that we have a built-in excuse to move to a new state. School/work is a great reason to do so. What's more, the process itself can be a lot of fun, and by the end of it, we all have a list of stories to share with one another about our travels.

Wearing Multiple Hats

As you have undoubtedly noticed from the prior chapters, neuropsychologists acquire many different skills through the training process, and this allows each one of us to potentially perform many different activities on a daily basis (e.g., psychotherapy, grant writing, cognitive testing, administrative duties, teaching). Indeed, we both find this to be one of the appealing aspects of the profession. On the other hand, it can be challenging because each of the hats that we wear could represent an entire career. For example, teaching is a demanding job, and some people spend their entire professional lives working toward improving their skills as an educator. There are even entire academic journals dedicated to this arena (*Teaching of Psychology*, *Contemporary Educational Psychology*, and *Scholarship of Teaching and Learning*). Consequently, if we spend 60% of our time conducting research, 30% of our time providing care to patients,[3] and 10% of our professional time teaching and supervising trainees, we don't have many hours left in the day to study pedagogy and improve our skills as educators.

If we are not careful, we can begin to feel like we are being pulled in different directions, with too much information and research to stay abreast of. Our coat closet full of different hats can lead to chaos and disorganization, especially if we

---

[2] It is exceedingly rare, but some people have managed to stay in the same geographic location for the entirety of their training (undergraduate, graduate school, residency, fellowship); however, this significantly restricted their options, making the application cycles more difficult (see Part II).

[3] Or, as one of our mentors and colleagues, Dr. Geoff Tremont, comically put it, "80% research and 80% clinical."

Table 4.2 Organizational advice

1) Become acquainted with your personal work habits (e.g., What time of day are you most productive? In what setting are you most productive?) and then take advantage of them.
2) Use a calendar religiously.
3) Minimize time spent commuting to work and use that time to be productive, if possible. For example, if you take the train, use your laptop; if you drive, listen to relevant lectures, podcasts, or books on tape.
4) Learn to say "No" when appropriate.
5) Schedule writing time into your daily work week.
6) Check out the *WorkLife* podcast with Dr. Adam Grant: https://www.ted.com/podcasts/worklife.

also have personal/social lives to attend to (and this is important, so we're told). You could find yourself hustling to catch up on 15 agenda items each day, while more research grants, clinical reports, and emails continue to pile up. This is likely to lead to stress and unhappiness. While you are not currently reading a self-help book and we do not purport to have all of the answers, Table 4.2 provides some tips to help you stay structured, organized, and content in your work life.

Even with excellent time management techniques, it is still possible to become overwhelmed with work and to experience "professional burnout" (Marek et al. 2017). This is not good for anyone – yourself, your family, your patients, your coworkers – and it is a very important problem that can be prevented through proper self-care. We will discuss this issue in further detail at the end of Ch. 6.

> Some of the best advice I ever got that I totally ignored was to say 'no.' There's only so much science one can do in a day and it's too easy as a trainee to come in and be so excited about everything and to say 'yes' to everything. It's all really cool and exciting but at a certain point you have to be able to say 'no' in order to keep your sanity. So, say 'no' to certain things and then use that time you gain to focus in on the tasks that are making the greatest impact on your career.
>
> – Adam J. Woods, PhD

Finally, we will mention that there are a few uncommon exceptions to the "wearing multiple hats" rule. In other words, it is possible to specialize narrowly in neuropsychology, finding one particular aspect of the profession that you thoroughly enjoy and spending the majority of your time in this domain. For example, some neuropsychologists at hospitals and in private practice settings do exclusively clinical work with a specific population. On the other hand, some research neuropsychologists are not interested in pursuing clinical work or teaching, so they focus solely on their scientific pursuits. There is no right or wrong answer here – it is all about what provides you with the greatest level of fulfillment, productivity, and work-life balance. Fortunately, as we have already mentioned, neuropsychologists are in high demand all over the country, so you are likely to find a job that suits your particular interests if you look hard enough and have some flexibility in terms of location and setting.

Ambiguity

Many of us crave confidence and clarity in our lives. As we mentioned above, there is certainly ambiguity in the training process, and once you have a stable job, you have a different type of ambiguity to contend with. Specifically, when you conduct a neuropsychological evaluation for a patient, you are often attempting to synthesize a large amount of complex information into a coherent, accurate, and helpful story for the patient, their family, and their doctor. Unfortunately, there is not always a clear diagnosis or conclusion to provide, and this requires us to be nuanced and communicate a lack of full certainty to our patients. For example, a 71-year-old woman may present to our office with a referral from her neurologist, wondering if the memory loss that she has noticed is due to Alzheimer's disease or if it is simply the result of normal aging processes. After a neuropsychological evaluation, we may find that she has consumed alcohol daily for years, that she struggles with depression at times, and that she takes multiple medications with the potential to impact cognition. Moreover, her testing may reveal several inefficiencies in attention, memory, and problem-solving compared to other women her age. In this case, we will probably not be confident one way or the other – does she have early-stage Alzheimer's disease? We don't quite know. It could be that the subtle memory loss that she reports and that we measured on our tests reflects the first symptoms of Alzheimer's disease, but it also could be due to long-term alcohol use, low mood, and/or polypharmacy. When we meet with her to share our results, we will need to explain what we found in a manner that is both understandable and representative of the complex reality.

Many of our patients are looking for a concrete, conclusive statement as to their cognitive and emotional health, but it is our job to be as accurate as we can, and the world is complicated. We would be doing our patients a disservice if we portrayed overconfidence about our findings and misled them in one direction or another. With our hypothetical older woman, we could present our findings to her as clearly as possible and then recommend that she (1) engage in evidence-based intervention for alcohol use, (2) receive treatment for her depression, (3) talk to her physician about the medications that could be negatively impacting her memory, and (4) repeat testing in a year or so in order to track any cognitive changes and clarify the etiology. Our hope is that, in lieu of a definitive diagnosis, these recommendations will help her improve her quality of life and daily functioning. What's more, if she follows our advice, it may allow us to better determine the cause of her problems. Specifically, if treatment for depression and medication alterations lead to a significant improvement in her memory, back to her personal baseline, then it is far less likely that she has Alzheimer's disease. On the other hand, if her memory issues persist in the face of good interventions and she comes back for follow-up neuropsychological testing 1 year later, showing even more decline and impairment, Alzheimer's disease would be more likely. In this case, our patience and ability to withstand diagnostic uncertainty allowed us to help our patient as much as possible, regardless of whether our ultimate conclusions are good news or bad news.

Similar to clinical work, there is ambiguity in research as well. This is the case in all of psychology, where our results can differ when we replicate prior work and where our data can be interpreted in multiple ways. Let's take an example to make this clearer. Imagine a simple chemistry experiment in which you light a fire and then add strontium nitrate (red) or copper chloride (blue) to change its color. If you repeat this experiment with the same ingredients, under the exact same conditions, you should obtain the same results each and every time. In psychology, this is not the case. You could run an experiment on a group of 60 people and find that, for example, eating whole grain cereal every morning improves the group's memory. I could then attempt to reproduce your results. I could follow the procedures of your experiment down to the most precise level possible, and I could still obtain different results. This is because human beings are complex! There is a high degree of variability in human genetic makeup, biochemistry, environment, and behavior. If my results are different than yours, it could simply be because I recruited a different sample of research participants than you did, and the interindividual (across individuals) differences mean that your results do not generalize to my experiment.[4] Now, this is not to say that research in psychology is meaningless. On the contrary, it is vitally important and meaningful. But it does mean that we have a difficult job to do, which is to make sense out of the complex, chaotic world that is human thinking and behavior. We must be careful, diligent, and precise, and we must show grit and persistence. We will undoubtedly face unpredictability in our work time and time again. Being aware of this on the front end can go a long way in terms of improving your mindset, your approach to research, and, ultimately, your success as a scientist.

****

In this section, we have discussed examples of ambiguity in neuropsychology. Now we will describe a few additional challenges specific to academic careers in neuropsychology and then follow up with challenges related to clinical work.

Academic Challenges

One specific challenge to a scientific career is that of navigating academic politics and the "publish or perish" culture in some institutions. Universities, like other organizations, have a hierarchical structure, with people harboring varying motivations and incentives. If one wishes to assimilate into the culture and appease the higher-ups, there can sometimes be pressure to spend time attending unnecessary meetings, to work with colleagues with difficult personalities, and/or to offer authorship on a paper where it may not be deserved.

---

[4] Dr. Brian Nosek, a researcher at the University of Virginia in Charlottesville, has done great work in this area (see Nosek et al. 2015).

With regard to the "publish or perish" mindset, there is a limited amount of funding for researchers, and people who fund their own work by writing grants are in competition for limited resources. It can be stressful to work under grant deadlines, knowing that both the success of your research idea and your salary depend on being awarded the grant. Moreover, the competition for funds and prestige can lead to interinstitutional rivalries and unsupportive cultures.

With all of that said, we have both worked in academic institutions and have found there to be great potential for personal and professional fulfillment in these settings. The politics and pressure to receive grant funding and publications vary from place to place, so we always encourage new professionals who are looking for jobs to ask plenty of questions of their future colleagues and coworkers in order to learn about the culture of the institution. Do some investigative work to determine if this is the type of environment in which you would be happy and successful. Hard work early on can pay dividends later.

Clinical Challenges

Clinical jobs can also have pressure (to see patients) and politics (you are working with other people after all), but these forces tend to be less salient in this setting than they are in some research positions. In contrast, there are a few hurdles specific to clinical work. The most common of these, and the bane of many clinicians' existence, is navigating insurance reimbursement.[5] For neuropsychologists in a few select settings (e.g., VA hospitals), this is not an issue, but many of us in medical centers and group practices must advocate for ourselves by filling out paperwork and making phone calls in order to receive reimbursement for our work. Typically, step 1 for neuropsychologists is an application for credentialing with different insurance providers who cover patients in their region (e.g., Medicare, Blue Shield, Anthem Blue Cross). Occasionally, the providers deny the application, meaning that they will not pay for any services provided to patients with that coverage, and this limits the potential pool of patients that can be seen. If they do credential the provider, then step 1 is finished and the clinician is able to submit a *request for reimbursement* for any patient who has that coverage. Those claims are occasionally denied (meaning they won't pay, even though the clinician has already completed the work), and sometimes the clinician is required to negotiate with the insurance representatives because there was some minor error in the filing process. What's more, some insurance providers require authorization before testing is completed; otherwise, they will refuse to pay for the services. Additional downsides are that each insurance provider can set the reimbursement rate and limit the number of hours and types of cases for which the clinician can be reimbursed.

---

[5] The information in this paragraph is specific to US-based clinicians.

All of this complexity can be a major headache. Fortunately, most clinicians have paid staff members whose job duties include submitting and following up on insurance claims. This lightens the burden on us and allows us to focus more on doing what we enjoy and were trained to do – providing neuropsychological services to patients.

A second major challenge to clinical work is delivering bad news to patients. This can come in many different varieties. It could be revealing to an older woman and her husband that she has dementia and will continue to slowly decline until the end, or it could be telling a middle-aged person who was in a workplace-related accident and sustained a concussion that the data from their cognitive evaluation were not valid,[6] or it could be explaining to the parents of an 8-year-old girl that you are diagnosing their daughter with intellectual disability. As you may imagine, this type of news can be very difficult to hear, and it is our job to explain, clearly and empathically, the conclusion that we have come to, the evidence that led to our conclusion, and the recommendations that we have for the patient and their family. For example, imagine that the parents of our 8-year-old girl were in denial about her cognitive status. She had been falling behind in her classes, struggling to learn to read and do math, and having difficulties solving problems and making connections between concepts. Her teachers expressed concern to the parents, who helped tutor her after school in the evenings and on weekends. Unfortunately, it did not help very much. With the passage of time, they began thinking that she may have a learning disability, and this led them to bring her in for a neuropsychological evaluation. But when you tell them that she meets criteria for intellectual disability (the more well-known but outdated term for this is "mental retardation"), it would not be surprising for them to have a major emotional reaction. They might become tearful and state that they have never thought that their little girl was "stupid" or "retarded." They may even push back against your diagnosis, stating that you don't know what you are talking about.

Clearly, some of our patient interactions are very difficult. Many people would turn around and run away as fast as they could from situations like this. But neuropsychologists are trained for just this sort of situation. We gain experience working with people who are in acute distress and who may be guarded and/or reactive in the room with us. It is human to show emotions when faced with difficult circumstances, and we pride ourselves on our ability to both (a) explain and educate people in ways that are helpful and (b) show compassion and empathy in a manner that helps us form a connection with our patients.[7]

---

[6]As we noted in Ch. 1, there are methods for determining whether someone is putting forth enough effort on testing to produce valid, interpretable data.

[7]For more information on providing neuropsychological feedback, see www.NavNeuro.com/17 and www.NavNeuro.com/29.

Conclusion

So there you have it. Those are a few of the challenges of working as a neuropsychologist. It is not a walk in the park, but we think that all of these hurdles can be crossed and that the benefits are well worth it. Now that you are fully informed about the field and the potential challenges, let's turn our attention to helping you begin your neuropsychological journey with a roadmap to the finish line.

References

Marek, T., Schaufeli, W. B., & Maslach, C. (2017). *Professional burnout: Recent developments in theory and research*. Routledge.

Nosek, B. A., Aarts, A. A., Anderson, C. J., Anderson, J. E., Kappes, H. B., & Open Science Collaboration. (2015). Estimating the reproducibility of psychological science. *Science, 349*(6251).

Part II
A Neuropsychologist's Journey

Chapter 5

Undergraduate Training

> If there is one thing we could do as a field it is to start establishing a more prescriptive set of guidelines for training in neuropsychology.
>
> – Margaret Lanca, PhD

If you have made it this far in the book, we take it that you are serious about our field (and that you are a nerd, like us). With that in mind, this chapter is designed to guide you through your undergraduate years, maximizing the chance that you will be accepted into graduate school and creating a solid foundation for your upcoming neuropsychology training. Whether you are a high school student considering your options, a current college student navigating coursework, or a working adult contemplating a career change, we believe that this chapter will place you well ahead of the curve. In our experience, the majority of neuropsychology trainees did not learn about the field until their late undergraduate/early graduate training. Graduate programs in psychology can be quite competitive, and you will be at a significant advantage if you begin acquiring the requisite knowledge sooner.[1]

As you move through this chapter, you might get the impression that we are prescribing the one "right" way to navigate college, but this is not the case. There is a fair amount of variability in the path taken through one's undergraduate training. Here, we provide tips and advice that you can apply to your own individual situation and that will help you prepare for the graduate school application process. Broadly speaking, the most important objectives to accomplish at this stage are the following: (1) increase your knowledge about psychology and neuroscience, (2) build

---

[1] Although a portion of the information will be specific to the neuropsychology path, much of our advice will apply to success in undergraduate training more broadly.

© Springer Nature Switzerland AG 2021   79
J. A. Bellone, R. Van Patten, *Becoming a Neuropsychologist*,
https://doi.org/10.1007/978-3-030-63174-1_5

good relationships with multiple professors, (3) gain research experience, (4) earn good grades, and (5) enjoy yourself! In order to help you achieve these goals, we have organized this chapter into the following sections: *Completing coursework, Developing professional and clinical skills, Acquiring research experience, Managing finances,* and *Applying to graduate school.*

Completing Coursework

Core Courses

The most common question that we have received from high school and college students interested in neuropsychology is, "What should I major in?" Although a small number of programs (e.g., the University of Cincinnati) offer a neuropsychology major, the vast majority of people in our field earn a bachelor's degree in psychology.[2] We think that this is appropriate because cultivating a strong foundation in general psychological concepts is a major asset in graduate school and beyond. Additionally, keep in mind that you will eventually be applying to doctoral programs in *psychology*, not neuropsychology. Thus, we recommend that you major in psychology. If you choose to major in a different, related subject (e.g., neuroscience or biology), be sure to enroll in the core psychology courses required by the graduate programs to which you will apply.

If you are in the position of having already graduated from college with a non-psychology degree, but you wish to pursue a career in neuropsychology, we suggest contacting clinical psychology doctoral programs directly to ask about whether they will consider your application and what you can do to make yourself as competitive as possible. They may suggest that you complete additional college-level psychology courses or seek out psychology-related research and clinical experiences.[3] We also recommend asking representatives from these graduate programs whether a high score on the Psychology GRE would increase your chance of being accepted into their programs.[4] Finally, you could consider completing a master's degree in psychology before applying to doctoral programs, although there is nuance here (see below).

---

[2] As we laid out in Ch. 1, neuropsychology exists under the umbrella of psychology. As such, we are trained broadly as psychologists; for example, we adhere to the American Psychological Association's (APA) ethics code (APA 2017), we receive training in psychological treatments, and we earn doctoral degrees in clinical psychology.

[3] Post-baccalaureate ("postbac" or "post-bacc") programs are potentially a good fit in these situations because they provide opportunities for these activities (e.g., coursework, research, clinical experiences) in preparation for graduate school. You can find out more about these programs here: https://www.apa.org/ed/precollege/psn/2014/09/post-baccalaureate

[4] For information about the Psychology GRE see https://www.ets.org/gre/subject/about/content/psychology

There will likely be a core class regimen built into your psychology major, and you should obviously focus on what is required to graduate. Importantly, all of the core courses (see Table 5.1) will be beneficial to your career because theories and empirical findings from all corners of the psychological landscape are invaluable to neuropsychology. Let's briefly discuss an example. A neuropsychologist would do well to be intimately familiar with the work of Russian physiologist Ivan Pavlov, who discovered classical conditioning, the phenomenon whereby a once-neutral stimulus (e.g., a tone) is paired with a biologically relevant unconditioned stimulus (e.g., the smell of food) to ultimately elicit a similar response (salivation). Classical conditioning is the basis of many otherwise unexplained human experiences such as the acquisition of disorders of trauma and anxiety, the power of advertisements on consumer behavior, and the danger of overdosing from taking illicit drugs in new environments. It can be of great import to a neuropsychologist when, for example, he assesses a patient with an anxiety disorder or posttraumatic stress disorder; if the patient's psychopathology was originally generated through the mechanism of classical conditioning, the neuropsychologist will be better able to understand the patient's distress and recommend appropriate treatments (see Powell et al. 2016).

Now that we have provided you with some general advice, let's get specific. One course that we highly recommend you add to your schedule is *biopsychology*, which is sometimes called *psychobiology*, *biological psychology*, or *psychophysiology*. This is perhaps the most widely offered course that is similar to neuropsychology. The content focuses on the relationship between (1) a physical substrate (the brain), and (2) the mental activity (the mind) and overt action (behavior) it produces. If this is all sounding familiar, you may recall from Ch. 1 that neuropsychology is the study of brain-behavior relationships. Hence, neuropsychologists tend to be very enthusiastic about biopsychology. In fact, I (Ryan) found my way into neuropsychology through a biopsychology course at my alma mater, James Madison University. My professor, Dr. Jeff Dyche, set aside one class period to discuss careers related to biopsychology and he spent no more than 10–15 minutes of that session on the profession of neuropsychology… but that was more than enough to get me hooked!

Table 5.1 Sample of core psychology courses

Statistics for the social sciences	Sensation and perception
Psychological research methods	Biopsychology
Personality psychology	Cognitive psychology
Social psychology	Psychology of learning
Abnormal psychology	Testing/assessment
Developmental psychology	Psychology of cultural diversity

Supplemental Courses and Electives

Up to this point, we have provided you with two simple, straightforward recommendations – major in psychology and register for a biopsychology course. However, if you are looking to go beyond a single major and diversify your college coursework, you could double major or major/minor, in which case it would be beneficial for the second focus to be in a relevant arena such as neuroscience, biology, statistics, child development, English/journalism,[5] a foreign language,[6] or criminology. As we emphasized in Part I, neuropsychology is very broad and receives influence from many other areas of study. For example, a subset of neuropsychologists spend a great deal of their time performing the work of statisticians, while others frequently operate within the legal system, and still others acquire knowledge of neurobiology that nearly rivals their medical counterparts (e.g., neurologists, neuropathologists). A second area of study in college will add breadth to your knowledge and will help you to stand out when you throw your hat in the ring at a competitive graduate program.

It is typically a straightforward process to select and register for the classes that are required for a major or a minor. However, you will eventually be faced with a number of important and not-so-easy decisions related to selecting elective courses. A bit of general advice here is to identify a person (e.g., professor, counselor) who can provide you with individualized instruction and guidance on which courses to take and when, in order to ultimately meet your career goals. Ideally, this advisor will have some familiarity with the field of neuropsychology, including knowledge of various graduate programs and potential faculty members, so that they can provide you with tailored advice given your individual interests and skills.

****

We mentioned earlier that it is important to acquire a breadth of knowledge in and around psychology. Although the majority of your coursework should be at least loosely related to psychology, this is also your chance to branch out and try something novel. If a topic looks interesting to you, even if it has little to do with neuropsychology, we encourage you to consider taking it as an elective. The importance of diversifying your knowledgebase cannot be underestimated. In fact, journalist and political scientist Fareed Zakaria doubles down on this point in his 2015 book, *In Defense of a Liberal Education,* which we recommend.

In addition to broadening your academic scope, we encourage you not to shy away from challenging classes. While your friends may be taking basket weaving or ribbon dancing 101 to round out their schedules with easy credits, we encourage

---

[5] Writing skills are highly valued in neuropsychology.
[6] Multilingual neuropsychologists are in high demand.

you to challenge yourself with exposure to difficult topics and/or by signing up for classes with professors who are known to have high academic standards. For example, I (John) took cellular neurobiology as an elective in my senior year. It was no cakewalk, but I learned a great deal and I built a strong connection with the professor, who helped guide my decision to pursue neuropsychology.

A Note to High School Students

In high school, you will probably not have nearly as many options in your course selection as will be available to you in college. No worries! Earn good grades in the classes you are required to take and look forward to the freedom that college will provide. One class that may be available and is likely most relevant to neuropsychology is an advanced placement (AP) psychology course, so we suggest that you sign up for that if possible. Additionally, consider volunteer work in the life sciences, as this could set you apart in your college application. For example, you could volunteer to work with children with autism or with older adults in a skilled nursing facility. Check with teachers and administrative staff about specific opportunities at your school.

Action Steps:

 Major in psychology.

 Take a biopsychology course.

 Consider a second major or a minor in a related field.

 Identify an academic advisor with knowledge of neuropsychology and ask for guidance in selecting elective courses.

 Seek out challenging courses while also maximizing your grade point average (GPA).

Developing Professional and Clinical Skills

It would be narrow-minded to view college as merely a series of courses, with the assumption that you attend class, complete your homework, take tests, receive your grades, and then go on your merry way. In reality, there is plenty of room for growth outside of the classroom. Some examples of such opportunities – in no particular order – include (1) building relationships with professors, (2) acquiring clinical experiences, (3) attending conferences, (4) cultivating professional skills, and (5) working in research labs. We will cover the research portion later in this chapter; for now, we will say more about the first four tasks.

Building Relationships with Professors

It may seem obvious and easy to do, but you would be amazed at how few students approach their professors and begin conversations with them. Many university faculty members lament that no one attends their office hours until the day before the final exam, at which time they are bombarded with last-minute questions from anxious students who are ill-prepared because they slacked off for the entire semester/quarter. Don't be one of those students! Instead, our advice here is to prioritize making yourself known to your professors by developing a relationship with each of them. There are myriad benefits to doing so, including the following:

(a) Your teachers are experts in the class topic and one-on-one time with them will allow you to learn much more than you can from passively listening to lectures.
(b) Professors are the gateways into research labs, which are important to your future (see below).
(c) You will need professors to write letters of recommendation for graduate school applications.
(d) Networking with people in psychology and related fields can open doors that you didn't even know existed.

Importantly, this advice comes from more than just our own experiences. Indeed, one study found that quality interactions with faculty members were the largest contributor to students' perceived preparedness for graduate school (Huss et al. 2002).

We understand that the idea of approaching someone you don't know to begin a conversation can feel somewhat intimidating at first, especially with faculty members who are especially esteemed and successful, and/or who may come across as less than warm and fuzzy. But keep in mind that there is little downside to this tactic and a lot of potential upside; if you approach someone and they act aloof and curt, then you can move on and you have not lost anything. On the other hand, if the initial interaction goes well, you may be reaping the benefits of it for years to come.

Now that we have laid out the benefits, how exactly might you go about cultivating relationships with your professors? Ideally, you will initially introduce yourself after the first or second class and express your interest in the material. Feel free to ask a question or two as well, as long as it is not something that you could answer by consulting the syllabus. This will make you stand out among your peers and your professor will be more likely to remember you later. As we mentioned in Ch. 3, we have both taught courses ourselves, and we can attest to the positive impact of this strategy on a teacher. As the semester/quarter progresses, we suggest that you occasionally drop by their office hours to ask relevant questions, discuss the material, and/or talk about your career options.

Acquiring Clinical Experience

We recommend that you complete some form of clinical experience (sometimes called an "internship" or "practicum") during college. This means that you work, either for pay or as a volunteer, in a psychology or neuropsychology-related setting where patients are assessed and/or treated. This often occurs during the latter half of the college experience, but if it is possible to sign up during your freshman or sophomore year, then we recommend that you consider doing so. The type of work and the time commitment of a practicum can vary widely, from an hour or two per week to 10+ hours per week, so be mindful of what you're signing up for. Importantly, many universities build these practica into the academic curriculum, which allows you to earn credits while simultaneously gaining clinical experience.

In terms of locating potential clinical experiences at your university, we have three ideas to get you started. First, psychology departments often keep a running list of sites where prior students interned or volunteered. Second, you can ask your advisor or counselor, or any professor in the psychology department, about potential clinical internships. Finally, you can take it upon yourself to reach out to local neuropsychologists at nearby hospitals and/or private practices and ask to shadow them and help out in any way possible. Even if it means that you will initially be assigned to clerical work such as scanning and filing record forms, it is worth the effort in exchange for a glimpse of the day-to-day world of clinical neuropsychology.[7]

If you do not have access to neuropsychology-specific settings, it is perfectly acceptable to seek out clinical opportunities in the arenas of general psychology or social work. Examples of these include mental health centers, crisis hotlines, homeless shelters, substance use treatment facilities, and school districts. I (John) was involved in quality assurance for a local psychotherapy practice where I was tasked with flipping through therapy notes to ensure that they contained all necessary information and signatures. In this role, I was able to get a sense as to what types of patients the therapists were seeing and I had many interesting conversations with the clinicians during their lunch breaks. At the end of my internship, the lead psychologist wrote me a very nice letter of recommendation that helped me gain acceptance into my graduate program.

Attending Conferences

Professional conferences are gatherings of people with similar backgrounds and interests for the purpose of sharing ideas and networking with others. Conferences are not limited to science and medicine. For example, people who work in the areas

---

[7] But don't be shy! Ask your boss/supervisor if it is possible to increase your responsibilities.

of banking and finance hold annual conferences. In neuropsychology, there are several professional organizations that offer yearly meetings and these can be a gold mine for students interested in learning more about the field, practicing their presentation skills, and speaking with like-minded people. Typically, a conference is held in a different city each year and occurs over several days. An attendee's agenda might include attending seminars and symposia from experts in various areas, attending (and possibly presenting at) scientific poster sessions, mingling with friends and colleagues, and meeting new people whose professional work is of interest. For students who are new to the field, the poster session can be a particularly fruitful event because it represents a chance to practice presenting one's research to others. It is a stepping stone to the eventual tasks of (1) delivering formal scientific talks to rooms full of listeners and (2) publishing research papers in respected, peer-reviewed journals (see below). In poster sessions, researchers and students boil their projects down and present them on a large display (e.g., $48'' \times 36''$), with text and graphs describing the background, methods, results, and conclusions of the study. They hang their posters on large bulletin boards with dozens of other people in the same room, and everyone else meanders around the room, reading posters, and asking questions to the presenters.

> Go to conferences and talk to people. Science is social and the exchange of ideas is critical. We all have different personalities, and it is very important for trainees to interact with other people in our field in an academic and social context.
>
> – Adam Brickman, PhD

We encourage you to attend at least one professional conference (ideally in neuropsychology) during your undergraduate years, and the more conferences you attend, the better. Look for a meeting that is geographically close, or use this as an excuse to travel. Table 5.2 shows several options in neuropsychology, all of which offer discounted registration fees for students.[8,9]

Cultivating Professional Skills

To be a successful neuropsychologist, it is very useful to have strong writing, presentation, and interpersonal skills. Don't sweat it if you are not adept in these areas yet, because there are many ways to cultivate foundational abilities and remediate

---

[8] There is a cost for both membership to the organization and registration for the conference. In addition to reduced registration fees, many organizations offer travel scholarships to a select group of trainees to cover the cost of conference attendance.

[9] Feel free to attend other conferences (e.g., Society for Neuroscience [SfN], and/or the Association for Behavioral and Cognitive Therapies [ABCT]), but Table 5.2 lists the four that are most directly relevant to neuropsychology. There are also many niche conferences for specific disease processes (e.g., multiple sclerosis, sport concussion, Alzheimer's disease) that you can consider attending.

Table 5.2 Neuropsychological organizations with affiliated conferences

Organization	Conference website
International Neuropsychological Society (INS)	www.the-ins.org/meetings
National Association of Neuropsychology (NAN)	www.nanonline.org (click "continuing education" then "annual conference")
American Academy of Clinical Neuropsychology (AACN)	www.theaacn.org/annual-meeting
American Psychological Association (APA), Society for Clinical Neuropsychology (SCN)	https://convention.apa.org/future-conventions

areas of weakness during the college years. For example, you can take writing classes or sign up for tutoring in writing composition. You can gain exposure to public speaking by joining the debate team, by enrolling in communications classes, or by joining a local chapter of Toastmasters International (www.toastmasters.org).[10] And you can build interpersonal skills by participating in clubs or sports teams or by attending an improv class.

Teaching skills are another important area of development. In order to strengthen these abilities, there may be opportunities to serve as a teaching assistant (TA), particularly if you performed well in one of your courses. To do this, you can ask a professor during the middle to latter portion of the semester whether you could "TA" for her the next time she offers the class. This would provide you with valuable teaching experience and help you master the material.

Another broad recommendation, with multiple advantages, is participation in Psi Chi, the international honor society in psychology. One of the many benefits of Psi Chi membership is the opportunity to gain leadership experience by becoming a chapter officer (visit www.psichi.org/page/chapter_leader).[11] Lastly, do not shy away from extracurriculars or unique experiences such as studying abroad or learning a new language because these will round you out personally and professionally. For example, the ability to assess and treat Spanish-speaking patients is a highly sought after skill.

---

[10] *Exposure therapy* is a very powerful psychological tool that is used every day to help people with crippling fear and anxiety. In a nutshell, the idea is that exposing an anxious person to the anxiety-provoking stimulus for long periods of time (without a catastrophic result) will gradually reduce their anxiety. In other words, if you have public speaking anxiety, practice! Through repeated exposure to the act of presenting in public, you will reduce your anxiety and improve your performance.

[11] There are other ways to get involved in leadership positions within psychology as an undergraduate, such as by joining your university's psychology club, or formally starting the club if your school doesn't have one.

Action Steps:	
	Talk to your professors after class and/or attend their office hours.
	Ask your department for a list of internship sites and reach out to those that look interesting.
	Attend at least one neuropsychology conference.
	Take steps to improve your writing, public speaking, interpersonal, teaching, and leadership skills.

Acquiring Research Experience

In addition to coursework, professional development, and networking, the university environment is a great place within which to acquire research experience. Whether or not you see yourself as a full-time scientist down the road, we highly recommend that you join a lab as a research assistant (RA) and maximize your productivity in this area. The reasons for this are numerous and include the following: (a) gaining first-hand experience with the Scientific Method, (b) studying a subject in-depth, (c) building your curriculum vitae (CV; your academic résumé), (d) learning how to defend and advocate for your work, (e) building a strong relationship with a professor, and (f) networking with lab mates and other researchers (Petrella and Jung 2008). Importantly, the focus of the lab need not be strictly within the purview of neuropsychology as long as it is related to psychology or biology more broadly.[12] For example, during my undergraduate years, I (Ryan) worked as an RA in one research lab, studying exercise, substance use, and health behaviors within a behavioral framework. I also worked as an RA in another lab, investigating attribution theory and intergroup relations. While these topics are typically considered to be in the domains of health psychology and social psychology, respectively, the experience that I gained in the conceptualization, design, execution, and reporting of scientific projects was invaluable to my career.

So, how do you join a research lab? If you followed the advice we laid out above, you will have already been talking to your professors and networking with people at your university. Consequently, you will have an idea about what is available to you in the department in terms of areas of study, specific research projects, and RA openings. With regard to the timeline, keep your eyes open for research opportuni-

---

[12]That being said, it is a plus if there is a professor with a shared research interest and/or is a neuropsychologist because they will likely have more contacts in the field and have a better sense of pertinent graduate programs.

ties beginning on day one and throughout your four-year tenure as an undergraduate. While it is not essential that you join a lab in your first year, the earlier you
begin accruing experience, the better. So, we suggest that you visit your psychology
department's office early on and ask for a list of current research projects. We also
recommend that you perform an online search of psychology and neuroscience professors at your university and read about their research interests, grants, and recent
publications. Although it is fairly likely that you will be drawn to the research interests of at least one of the professors you meet in your classes, you wouldn't want to
miss out on a great opportunity in your department simply because you didn't happen to take a class taught by a particular faculty member.[13]

If you find a research lab that looks interesting to you, but you have not yet had
contact with the Principal Investigator (PI),[14] email the professor directly and ask if
they have an opening and would be interested in meeting with you. Here is an
example of a "cold" email:

Hi Dr. Martinez,

My name is [rock star student]. I am a sophomore at [XYZ university], majoring in psychology, and I came across your research on the effects of compensatory cognitive training on
functional outcomes in people with schizophrenia. This topic is particularly relevant to me
because I am interested in pursuing a career in adult neuropsychology with a focus in severe
mental illness. I would love to schedule a meeting with you if there is room for a research
assistant in your lab. I have attached my CV and I am happy to provide you with additional
information if it would be helpful.

Thank you for your time,

~[rock star student]

Similar to approaching a professor who doesn't know you and beginning a conversation, it can feel intimidating to cold email an esteemed researcher whom you
have never met. We are here to instill confidence in you. You have much to gain and
nothing to lose from sending out an email. We have been on both sides of emails like
the one we templated above, and we wholeheartedly encourage you to send out such
messages if there is a chance that you could connect with someone who could further your career.

When you do eventually schedule the first meeting with a potential research
advisor, treat it as an interview of sorts. Prepare ahead of time by browsing their

---

[13] There are additional research experiences that are open to many students. We recommend that
you look into The National Science Foundation (NSF) Research Experiences for Undergraduates
program (https://www.nsf.gov/crssprgm/reu/index.jsp), the Howard Hughes Undergraduate
Research Fellow Program (https://www.hhmi.org/science-education/programs), and APA's list of
options (https://www.apa.org/education/undergrad/research-opportunities)

[14] The leading researcher in a lab is commonly referred to as the Principal Investigator, or "PI."

online academic profile, searching their name on Google Scholar (https://scholar.google.com) and/or PubMed (www.ncbi.nlm.nih.gov/pubmed), and then reading a few of their recent publications. Think through why you would be interested in this line of research and how it is relevant to your future. It may also help to create a list of questions to ask the person. These questions serve the dual purposes of information gathering and demonstrating your interest and preparedness to them. In other words, most professors are impressed when a student has a list of relevant and thoughtful questions to ask them. And, finally, don't forget that this is a job interview – dress business casual or formal, depending on the situation.

****

Okay, now that you have established contact with professors, inquired about research labs, received invitations to labs, and nailed the interview(s), what's next? Well, after you join a research group, your experience will depend largely on the specifics of the work that you are engaged in. If you have done your due diligence on the front end and spent time learning about the lab in question, you will already have a general sense as to how your time will be spent. Either way, get ready to buckle down and enjoy your first foray into science!

Regardless of the specific lab you join, you will want to read up on the topic of interest (this process is often called performing a "literature review"). If you are not accustomed to consuming content from scientific journals, there will be a learning curve. Don't expect to breeze through it like you would a Harry Potter novel. Scientific articles are written for other scientists, not for the lay public, so there is a large amount of jargon and complex language. Don't let this discourage you. Follow the steps outlined in Table 5.3 when you begin consuming the scientific literature.

Before reading articles, you need to know where to find them (e.g., Google Scholar, PubMed). Unfortunately, some are not accessible without either a subscription to the journal or payment of a fee to purchase individual publications. Luckily, your university library will likely grant you access to those resources that you are not able to find on your own, so check with a librarian for help. As you begin accumulating more papers, we recommend creating an organized electronic storage system to keep them straight. For example, I (Ryan) use Dropbox software (www.dropbox.com) and I keep all of my papers saved into folders, organized by topic and sub-topic.

****

Table 5.3 A beginner's guide for reading a scientific article

1. Take your time, read slowly, re-read as necessary, and look up definitions of unfamiliar terms.
2. When content is highly dense and technical, don't get bogged down in the details; step back and try to understand the big picture.
3. Summarize in your own words what you have read for each section of the paper: a. introduction, b. methods, c. results, and d. discussion.

In the last few paragraphs, we have been discussing one very important skill: consuming the scientific literature. The next level to think about – and something that has huge benefits to a career in neuropsychology – is to publish your own work in well-respected, peer-reviewed journals. We have already outlined one stepping stone to such publications when we discussed poster presentations in the *Attending conferences* subsection. Now we will share that journal publications are the currency of the academic world. Graduate schools, internships, postdoctoral fellowships, and professional jobs all greatly value scientific productivity. So, we suggest talking to your lab's PI about whether it is feasible for you to serve as the first author or co-author[15] of several papers during your undergraduate years.[16] Tasks involved in contributing to these papers include the following: generating research questions, designing study methodology, carrying out study protocols, collecting data, analyzing data, and writing up results in the manuscript that will be submitted for publication. The details of honing scientific writing skills are complicated and beyond the scope of this book. For now, we will encourage you to check out the "Hourglass Method" of constructing manuscripts (Schulte, 2003) and we will refer you to several helpful resources:

• APA Publication Manual, 7th edition (APA 2019)

• APA Presenting your findings, 6th edition (Nicol and Pexman 2010a)

• APA Displaying Your Findings, 6th edition (Nicol and Pexman 2010b)

• A Short Guide to Writing About Psychology, 3rd edition (Dunn 2011)

Action Steps:

 Identify professors with whom you would like to do research.

- Talk to the faculty members who teach your classes.
- Search your department's website online.
- Ask your department if they have a list of ongoing research projects.

 Once you identify a lab, email the PI and ask if there is an opening for an RA.

 Present multiple posters at conferences.[a]

 Submit at least one article for peer-reviewed publication.[a]

[a]This should be done under the guidance of your research advisor

---

[15]Authors are typically listed in order of contribution. The person who contributes the most to the paper is listed first and everyone else is listed according to the amount of work they put in. Hence, being listed as the "first" or "lead" author is more valuable than being placed further down the list. One exception is that the "senior" author, or PI of the lab, is often listed last. Senior author publications are useful for experienced scientists who mentor younger researchers.

[16]Although adding publications to your CV is very valuable and makes you more competitive, this is not a requirement for graduate school.

Managing Finances

Financing college tuition can be a monumental task. Some people come from wealthy families who can cover tuition expenses, while others are forced to borrow tens of thousands of dollars in loans. This section is written with the second group of people in mind. First and foremost, we hope that cost does not prevent you from pursuing a higher education if that is what you want for yourself. However, we caution against blindly accepting student loans because it can take a tremendous amount of time and effort to pay them back, especially when factoring interest into the equation.[17] Fortunately, there are many other options available to reduce college tuition fees. We have divided this section into advice for high school students and advice for college students because the strategies vary depending on where you are at in your education. Here are some options to make college more affordable:

For those who are in high school:
- Take advanced placement (AP) classes and/or college-level examination program (CLEP) tests to earn college credit. Check to ensure that the universities you are interested in accept the credits. You can learn more about the CLEP program here: https://clep.collegeboard.org/school-policy-search.
- Enroll in classes at your local junior (i.e., community) college (called dual enrollment), and/or consider attending a junior college for the first 2 years and then transferring to a university to complete your degree.[18] You will receive the same diploma as students who attended the university for 4 years, and many universities automatically accept students from certain junior colleges if they meet the requirements (e.g., a 3.0 GPA). However, keep in mind that some people who attend community colleges are forced to remain enrolled at a university for one extra year (a fifth-year) because they were not able to acquire as much experience (e.g., research, clinical internships) as they needed for graduate school.
- Prioritize the academic strength of the college, not its fancy dorms or state-of-the-art gym. You do not need to attend a private/expensive school in order to gain acceptance into a reputable graduate school.
- Search for scholarships. You would be amazed at how many scholarships are available and how few people apply for them. If you have a solid GPA and perform well on the scholastic aptitude test (SAT) exam or the American college testing (ACT) exam, you will likely be eligible for various merit awards. Many scholarships have similar themes, so you can often use one essay (with minor tweaks) for multiple scholarship applications.
- If your family income is low, consider need-based financial aid. This type of assistance comes with fewer strings attached than non-need-based financial aid and includes options such as the Federal Pell Grant, direct subsidized loans, and the federal work-study program.

---

[17] For example, paying $500 per month toward a $50,000 loan at a 5% annual fixed interest rate would take over 10 years to pay off and would cost you an extra $15,000 in interest (so the total bill would be ~$65,000). You can do your own calculation here: https://www.finaid.org/calculators/loanpayments.phtml; https://studentloanhero.com/calculators/

[18] The reason this strategy saves you money is because attaining college credits at the cheaper junior college rate allows you to bypass those classes at a university and thus not pay the higher tuition fees.

For those who are in college:

- Search for scholarships.[19]

- Work part-time.
  - Any job can help pay the bills, and some companies offer tuition assistance in addition to a base salary.
  - Working can help build communication abilities, problem-solving skills, and overall self-confidence.
  - Consider jobs that allow you to network and/or gain neuropsychological experience (e.g., see the *Acquiring clinical experience* sub-section covered earlier).
  - Consider jobs that include free time to work on homework while on the clock (e.g., attendant at a library or recreation center, cashier). One example of this is a work/study program, which is offered by many colleges.

- Serve as a resident assistant in an on-campus dorm. Resident assistants are often provided with free housing in addition to other perks such as access to pre-registration (so you can take the classes you want, when you want).

- Serve as a TA and/or a tutor. TAs and tutors are sometimes provided with hourly pay and/or tuition assistance.

- Consider tuition assistance through military service and/or reserve officers' training corps (ROTC). If you have already served in the military, then use your GI bill.

- If you do borrow money, try to avoid loans with high interest rates. Also, minimize unsubsidized loans.[20]
  - Unsubsidized loans begin accruing interest as soon as you receive the money, even if you don't have to pay anything yet. That means that the amount you owe will continue rising at a certain percentage, making the grand total much higher than the amount you received.
  - Subsidized loans do not accrue interest until after the deferment period; in some cases, you can defer them as far as the completion of your postdoctoral fellowship.

- Read blogs about how to "hack" college in a variety of creative ways. For example, check out the college investor (www.thecollegeinvestor.com) and the resources we list in Ch. 7.

- Set yourself up for financial success in graduate school by excelling in college. The more successful you are in your undergraduate years, the better your chances of acceptance into a graduate program that will cover your tuition and potentially even pay you a stipend.

- When applying to graduate programs, consider asking them the following questions:
  - Do you provide tuition remission, health insurance, assistantships (research or clinical), a stipend, or other forms of funding?
  - How many students are funded?
  - For how many years are students typically funded?
  - What are the stipends and how much tuition remission is provided?
  - How do you determine who receives funding and how they maintain it?
  - How much debt do students typically have by the end of the program?[21]

---

[19] In addition to the multitude of public and private scholarships available, Psi Chi has a list of undergraduate psychology awards and grants: https://www.psichi.org/page/awards#undergraduate

[20] The following resources are packed with information regarding loans: https://tinyurl.com/yy8g-bweu; https://www.apa.org/monitor/2016/04/cover-debt-trap; https://www.consumerfinance.gov/paying-for-college/

[21] One study conducted by the APA found that the median graduate debt for early career professionals who had completed a clinical psychology program was $98,000 (Doran et al. 2016), which should act as your motivation to seek out those forms of non-loan financial assistance.

Action Steps:

🧠 Apply for as many scholarships and/or awards as possible.

🧠 Work a part-time job related to psychology.

🧠 Serve as a resident assistant, TA, and/or tutor.

🧠 Ask prospective graduate programs important financial questions.

Applying to Graduate School

Introduction

As you learned in Part 1, the standard path to becoming a neuropsychologist involves attending graduate school and obtaining a doctorate in psychology. Thus, an important task in college is applying to graduate programs, which is why we devote the remainder of the chapter to this process. Following the steps that we laid out above (e.g., completing required coursework, obtaining research experience) will ensure that you are poised for success in this competitive endeavor. In addition, we recommend that you read about the graduate school application process and requirements. There are several books that are devoted to this topic. If you were to read just one of them, we highly recommend the most recent edition of *Insider's Guide to Graduate Programs in Clinical and Counseling Psychology* (2020) by Michael Sayette and John Norcross.[22] Because Drs. Sayette and Norcross cover all aspects of the process in great detail, we will provide only the highlights here, and we will make the information specific to neuropsychology where appropriate. As you move through this section, keep in mind that applying to graduate school is quite different from applying to college for multiple reasons, the most notable of which is that you are applying to a specific program (e.g., The Clinical Psychology Program at the University of Washington) rather than a university more broadly.

---

[22] Others include the following: (1) Psi Chi's *An Eye on Graduate School: Guidance Through a Successful Application*, (2) APA's *Graduate Study in Psychology*, (3) *Applying to Graduate School in Psychology: Advice from Successful Students and Prominent Psychologists*, and (4) *Getting in: A Step-By-Step Plan for Gaining Admission to Graduate School in Psychology*.

Application Timeline

It is never too early to begin thinking about graduate programs and preparing materials, but we recommend that you begin ramping up the process over 1 year prior to the first day of the graduate program. For undergraduate students expecting to proceed directly to graduate school, this means that you begin the application process in earnest during the early to middle portion of the summer before your senior year. This approach will allow you ample time to thoroughly investigate the programs, take the Graduate Record Examination (GRE), acquire letters of recommendation, and complete the applications, which are typically due in December. If you choose this route, you will benefit from planning ahead by taking fewer courses and cutting back on working hours during the Fall quarter/semester of your senior year. Keep in mind that the application process is a part-time job in and of itself.

Many people choose to postpone graduate school after finishing college.[23] There are several reasons for this, including (a) a late start on the application process, (b) an application that is underdeveloped relative to the applicant's graduate programs of interest, and (c) uncertainty about one's career path. Of course, you will need to decide on the best time to apply given your unique situation. Regardless of when this might be, Table 5.4 can help you get started with a rough timeline of the application preparation process. We will cover each of these milestones in more detail.

Identifying Programs

Choosing graduate programs to apply to can be both exciting and daunting. As of early 2020, there were 395 active APA-accredited doctoral programs (https:// accreditation.apa.org),[24] so it is natural to feel overwhelmed and unsure about how to proceed. There is no right or wrong way to go about this, but we have some general guidelines and resources to share with you.

First, if you are geographically restricted, then you will obviously have a much smaller pool from which to choose. Depending on where you live, this could severely limit your opportunities. Although we understand that some people are truly inflexible when it comes to relocating, we urge you to consider whether there is any wiggle room for expanding your geographic radius. The larger this region is, the greater your chances of acceptance into a program that fits with your interests and career goals.

---

[23] One study sampled clinical psychology PhD students and found that 33% went straight into their program upon graduation from college, while 57% postponed beginning graduate school; the remaining 10% proceeded to a terminal masters' program before eventually attending their PhD program (Zimak et al. 2011).

[24] The APA only accredits programs in the US and its territories. For Canadian readers, visit the Canadian Psychological Association (CPA) website: https://cpa.ca/accreditation/cpaaccreditedprograms/

Table 5.4 Sample application timeline with preparation steps

Amount of time prior to your application deadline	Preparation step
8+ months	Begin studying for the GRE general test
~7 months	Begin researching programs and creating an excel sheet to organize the information (you can find an example at www.NavNeuro.com/book)
	Take the GRE for the first time
~6 months	Begin studying for the GRE psychology subject test (if needed)
~5 months	Begin drafting your personal statements/cover letters and ask friends/family to read them
~4 months	Re-take the GRE (if needed)
	Narrow down the number of programs to which you will apply.
	Take the GRE psychology subject test (if needed); note that this is only offered in September, October, and April, and you must register ahead of time
~3 months	Ask professors if they are willing to write strong letters of recommendation for you
	Ask several close faculty members to review your personal statement and CV (after updating it)
~2 months	Finalize the list of the ~12 programs that you will be applying to.
	Forward GRE scores and transcripts to the graduate programs
1 week (or sooner)	Submit your application and confirm that the program has received it

Other than geography, the answers to the following questions will help narrow the field:

Must the program be APA accredited? The short answer is, "Yes." There are many reasons for this, but the most compelling argument is that attending a non-APA-accredited program jeopardizes your ability to obtain a doctoral internship, postdoctoral fellowship, state licensure, board certification, and certain jobs.[25] These are all of the important steps in becoming a neuropsychologist. Keep in mind that the specific *program* (i.e., clinical psychology PhD) needs to be APA accredited (or "accredited, on contingency"[26]), not the university. To reiterate, if you are looking to attend graduate school in the US, we strongly suggest that you only consider APA-accredited programs.[27]

Should I obtain a PhD or a PsyD? When choosing to pursue psychology doctoral programs, there are two typical degree choices in the US: PhD (Doctor of Philosophy)

---

[25] For additional information, visit https://accreditation.apa.org

[26] This means that the program is new and does not yet have outcome data to submit to the APA. This counts as APA accredited for your purposes.

[27] Similarly, if you are looking to attend graduate school in Canada, we strongly suggest that you only consider CPA-accredited programs.

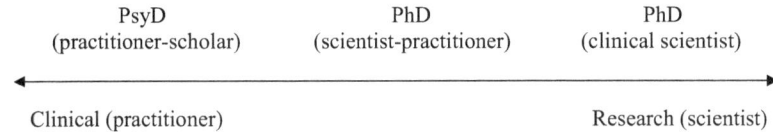

PsyD	PhD	PhD
(practitioner-scholar)	(scientist-practitioner)	(clinical scientist)

Clinical (practitioner)   Research (scientist)

Fig. 5.1 Spectrum of models, from clinically focused to research-focused

or PsyD (Doctor of Psychology).[28] Contrary to how it sounds, pursuing your PhD does not mean that you will be required to read Aristotle's full treatises in metaphysics and Nicomachean ethics (although we still recommend it!). You will work toward a PhD *in clinical psychology*. Both the PhD and PsyD provide training in clinical practice, but a PhD includes a stronger research emphasis than a PsyD. Most programs that offer a PhD in clinical psychology adhere to the "scientist-practitioner" (aka "Boulder") model, meaning that they prepare students both as clinical practitioners and as scientists, whereas PsyD programs adhere to a "practitioner-scholar" (aka "Vail") model, where they prepare students to be practitioners. A third, more recently developed model offered by some PhD programs is a "clinical scientist" model, which is further down the "scientist" side of the continuum and focuses on training academics rather than clinicians. Figure 5.1 depicts this spectrum of models, with the scientist-practitioner PhD model squarely in the middle of the other two.

Aside from the research focus, there are several other general differences between PsyD and PhD programs. Typically, PsyD programs:

(a) Are one year shorter than PhD programs because there is less research training and typically not a formal dissertation
(b) Accept more students, so they have larger class sizes and less individual attention from professors
(c) Have higher acceptance rates
(d) Offer less financial assistance, meaning overall debt burden is higher[29]

In terms of outcomes, PsyD students tend to have lower match rates to accredited internship programs (APPIC 2019) and lower licensing examination scores (Schaffer et al. 2012). Two caveats are important here: (1) these are merely averages and there is a great deal of variability, both by individual and by program, and (2) the majority

---

[28] One other degree is a Doctor of Education (EdD), which is a professional doctoral degree in education. There are a few neuropsychologists with an EdD, but they are in the very small minority. Going this route typically requires substantial re-specialization to ensure the acquisition of requisite knowledge/skills (see Ch. 7). Also, the APA does not accredit EdD programs, which severely limits neuropsychological opportunities for these degree holders.

[29] One study conducted by the APA found that the median final graduate debt for PsyDs was $138,500 compared to $67,000 for PhDs (Doran et al. 2016). Another study found that 40% of PsyDs had over $200,000 of post-graduate debt compared to only 11% of PhDs, and that 41% of PhDs had less than $25,000 of debt compared to only 10% of PsyDs (Whiteside et al. 2016).

of PsyD students match to accredited internships and pass the licensing exam on the first try.

So, the obvious question is, "Which degree should I choose?" We'll make it easy for you, and say that we recommend that you pursue a PhD.[30] Even if you want to spend 100% of your time as a clinician (i.e., 0% as a researcher), the PhD route offers more flexibility and research training that will be attractive to doctoral internship and postdoctoral programs (Driskell et al. 2020; Ritchie et al. 2012). A widely endorsed and well-justified notion is that scientific training enhances clinical practice (Baker and Benjamin 2000), so the research experience provided by PhD programs is useful to academics and clinicians alike. Furthermore, the Houston Conference guidelines policy statement notes that, "the training of the specialist in clinical neuropsychology must be scientist-practitioner based" (Hannay et al. 1998, p. 1). We do not believe that this means to exclude individuals from practitioner-scholar model programs,[31] but rather to emphasize the importance of research training. Finally, the PsyD degree is less well-known to patients, physicians, and other healthcare workers, so you may find yourself explaining what a "PsyD" means and justifying why you didn't obtain a PhD. Relatedly, and unfortunately, the majority of PsyD students pursuing neuropsychology have experienced bias associated with their degree (Whiteside et al. 2016).

If you do take the PsyD route, it is still absolutely possible to become a neuropsychologist, and many outstanding clinicians and researchers have PsyDs. In the studies we referenced in the last paragraph (Driskell et al. 2020; Ritchie et al. 2012), although most postdoctoral training directors said that they prefer PhD over PsyD applicants, nearly all indicated that a PsyD is acceptable. Also, rates of PsyD degrees among neuropsychologists have increased over the years (Rabin et al. 2016; Sweet et al. 2020), and we hope that the perceived disadvantage will decrease over time (and, as a result, our recommendation could change). Importantly, there are ways to minimize the potential discrepancy in outcomes between the two degrees. Because PsyD programs are considerably heterogeneous (i.e., the quality of training varies substantially; Norcross et al. 2004), the process of vetting graduate programs is very important.[32] We also recommend that you engage in a significant amount of research during graduate school, obtaining publications, and delivering as many presentations as possible. This likely means that you will go above and beyond your program's requirements. You may even consider taking an additional year to complete your doctorate, allowing you to gain as much research and clinical experience as your PhD counterparts whose programs are typically a year longer than PsyD programs.

---

[30] Keep in mind that we both have PhDs and we are not free from bias.

[31] Neuropsychology boards accept both PhD and PsyD degrees.

[32] Generally speaking, programs that are embedded within a university's psychology department offer more resources and well-rounded training than freestanding programs.

Should I aim for scientist-practitioner models or clinical scientist models? This is a complex question and people have varying opinions about it. We think that both models have a great deal to offer and some people are best suited to one over the other. When considering scientist-practitioner versus clinical scientist, keep in mind that the former is more clinically oriented (think, scientist-*practitioner*), and the latter is more research-focused (think, clinical *scientist*). Consequently, if you are very confident that you want a clinically oriented career, scientist-practitioner programs will be the best fit, while if you feel strongly that you want to spend the majority of your time in research, the clinical scientist model is right for you.[33] Importantly, many people don't know whether they will lean toward clinical work or research; in that case, keep in mind three important facts. First, the scientist-practitioner model is typically more balanced than the clinical scientist model – it includes more of the secondary focus (research) than does the clinical scientist model (clinical work).[34] Second, it is typically easier to transition from a research career to a clinical career than it is to transition from a clinical career to a research career. Third, you can apply to both types of programs, and many students do this. Given that acceptance rates are lower for research-oriented programs (Norcross et al. 2010), those who apply to these programs would do well to diversify their portfolios with a few scientist-practitioner program applications in order to maximize the likelihood that they will be accepted into graduate school.

Should I choose a clinical or counseling psychology program? When you begin researching programs, you'll notice that they specify whether they specialize in clinical or counseling psychology. These subfields are less distinct than they used to be, but there are still important differences that lead us to recommend a *clinical* psychology program:

- Clinical programs outnumber counseling programs by a large amount (247 vs. 76, per 2018 data), so training directors/employers will be more familiar with the clinical designation.
- Counseling programs focus more on vocational (career) related assessment, whereas clinical programs are more likely to focus on cognitive assessment, which is the bread and butter of neuropsychologists.
- Very few clinical programs require a master's degree, while many counseling programs do (Norcross et al. 2014).
- Trainees from counseling programs sometimes do not meet institutional credentialing requirements for neuropsychological fellowships.

---

[33] One study found that over 75% of graduates from clinical scientist model programs are employed in academic positions or mixed academic-practice positions, relative to ~50% from scientist-practitioner model programs and ~17% practitioner-scholar model programs (Ready & Santorelli, 2014).

[34] In addition to APA, there are two main accrediting bodies for clinical scientist programs: Psychological Clinical Science Accreditation System (PCSAS) and Academy of Psychological Clinical Science (APCS). APA accreditation is important for those who want to pursue a career that involves clinical work.

- Most importantly for our purposes, neuropsychology is typically a focus of faculty in clinical (not counseling) programs. Indeed, at the time of writing this book, all of the programs we found that listed a neuropsychology concentration/track were clinical psychology programs.

How do I maximize fit? Fit is a very important concept that will come up multiple times throughout your training as a neuropsychologist. We always want to match students with programs that are right for them in terms of strengths, areas of growth, and career goals. Moreover, fit is bidirectional, meaning that you as the student are right for the program and vice versa. To ensure that you can contribute to a particular program and research lab, search potential faculty members online and read about their current students and ongoing projects. If helpful, reach out to professors to inquire about whether they will be taking graduate students and express your interest in applying. On the other hand, to make sure that a particular program is a good fit for you, consider your own interests and goals. If you are looking for a particular graduate school experience (e.g., pediatric neuropsychology, forensics), then find those programs that offer training in this area. Remember that the goal is not merely to get into a graduate program, but rather to get into the program that will prepare you to accomplish your career goals.

How important is my research advisor? In a word, *very*. Because of the importance of this person to the rest of your training and career, we urge you to think carefully about the quality of professors at prospective programs. First, consider the potential research mentor's productivity – that is, the number and significance of their funded grants and peer-reviewed publications (found in their CV). The more productive they have been in the past, the more productive they are likely to be in the future, and you can ride their coattails to further accomplishment and success (while also working hard, of course). Second, the personality fit between mentor and mentee is critical to the relationship. You will be spending a great deal of time together, and you want someone with whom you can communicate openly and honestly, while enjoying the process. In order to determine the personality fit, pay close attention to the interpersonal connection during your interview with them. Ask yourself how you felt when you talked with them – this includes both "shop talk" and "small talk." You can also ask them directly about their mentoring style. In addition to your own initial impressions, you can rely on other students' comments and suggestions. Other students might be particularly informative as it applies to more objective qualities such as responsiveness and availability, as opposed to a personality fit, which varies from person to person.

＊＊＊＊

To recap, although a PsyD or counseling PhD could still offer you the opportunity to pursue a career in neuropsychology, we are advising that you only consider APA or CPA-accredited Clinical Psychology PhD programs.[35] That narrows it down somewhat, but there are still a large number of programs to choose from: 200+ at last count (https://www.accreditation.apa.org). Out of these programs, only a subset have a strong neuropsychology presence, meaning that (a) they have professors who specialize in neuropsychology and (b) they have a strong track record of producing neuropsychologists. We believe that choosing a program with these characteristics is very important because the graduate school years are a prime time to begin receiving specialized neuropsychological training.[36] Consistent with this recommendation, training directors at both the internship and postdoctoral fellowship levels have indicated that it is important for doctoral students to receive neuropsychological training (Driskell et al. 2020; Ritchie et al. 2012).

We mentioned earlier that some programs offer a neuropsychology concentration or track. These are the programs that we recommend you consider first (you can find an updated list in the *Insider's Guide*). However, we also recommend broadening your search to all programs that have strong research and clinical opportunities in neuropsychology because there are some outstanding programs that do not offer a formal concentration but still produce very well-trained neuropsychologists. The *Insider's Guide* also includes these in the appendix.[37] Furthermore, you can ask programs whether they offer neuropsychological curriculum/training that adheres to the Houston Conference guidelines (see Ch. 6 for details), as well as how many of their students have gone on to obtain board certification in clinical neuropsychology.[38]

****

From here on out, the decision about where to apply is contingent on idiosyncratic factors. For example, you may alter the programs to which you apply based on your GPA and GRE scores, using the aforementioned metrics as a barometer of your chances at acceptance.[39] Also, depending on your financial situation, you may

---

[35] At first glance, it may appear that we are stifling diversity by providing this recommendation. However, there is ample room for flexibility and variability in training from here.

[36] To be clear, we feel strongly that it is important to receive both generalist training in psychology and training specific to neuropsychology.

[37] The 2020/2021 edition of the *Insider's Guide* has research areas listed in Appendix E, specialty clinics, and practica sites in Appendix F, and concentrations/tracks in Appendix G.

[38] Some programs have begun specifying that they offer neuropsychology as a "Major Area of Study," which is the language specified in the Taxonomy for Education and Training in Clinical Neuropsychology (Smith and CNS 2019). As such, you may want to seek out these programs and ask all prospective sites where they fall on the Taxonomy (see Ch. 6).

prioritize programs that offer full or partial financial assistance. Some programs cover tuition and pay stipends to students; typically, these programs are more research-heavy and competitive, and you are more likely to be accepted if you have above-average research productivity and stellar GPA and GRE scores.[40]

> When I was applying to graduate school, my mentor encouraged me to look at programs where I would receive financial funding. This didn't seem to be really important to me at the time because paying back loans seemed so far away and I thought, 'well, if I take loans out, I will be able to pay them back later – I will be a neuropsychologist, it will be fine.' But I took his advice and I went to a program where my tuition was covered and it made a huge difference to me. So, if you have the opportunity to do this, it will positively impact you for the rest of your life.
>
> – Christine Koterba, PhD, ABPP-CN

One common question is, "How many programs should I apply to?" Generally, we encourage you to apply to about a dozen programs.[41] This number may seem large, but graduate school is 4–6 years of your life and these are formative years in terms of your professional training. Consequently, several additional applications are well worth the effort. However, although we recommend 12 as a benchmark, the ultimate decision depends on a number of individual factors such as geographical restrictions, GPA, GRE scores, and your anxiety level/risk tolerance. One helpful strategy is to include several programs where you are "shooting for the stars" and at least a few programs that are "sure bets" (i.e., you exceed their requirements and averages).[42]

Getting Organized

Now that you have an idea about exactly what you are looking for, it is incredibly important that you organize your thoughts and preferences. Create a spreadsheet that lists each program and then fill in relevant data: application deadlines, requirements, geographic location, contact person, presence/absence of a neuropsychology concentration, etc. We created a sample Microsoft Excel spreadsheet that is available for download at www.NavNeuro.com/book. Once you're organized, you can begin looking at the statistics and requirements of specific programs. There are several resources available that provide detailed information in this regard. In addition to examining the program's website to answer your questions, the *Insider's Guide*

---

[39] One study found that the mean overall undergraduate GPA among students accepted into clinical psychology PhD and PsyD programs in 2013 was 3.63 ($SD = 0.15$; Norcross, Sayette, & Pomerantz, 2017). The same study found a mean GRE Psychology Subject Test score of 686 ($SD = 39$), mean GRE Verbal Reasoning score of 159, and mean GRE Quantitative Reasoning score of 152. You should check the data from the graduate programs in which you are interested.

[40] See the *Managing finances* section above regarding important financial questions to ask prospective programs.

[41] Psychology doctoral applicants apply to ~10 programs on average. We have seen a range of recommendations, from as few as 6 to as many as 20.

[42] See the *Insider's Guide* for acceptance rates by program.

Table 5.5 Websites that identify clinical psychology programs

Website	Description
https://scn40.org/ training-directory/	This society for clinical neuropsychology (SCN) site includes a search engine of both American and Canadian programs with a neuropsychology focus. This is not an exhaustive list, but it serves as an excellent starting point.
https://accreditation. apa.org/ accredited-programs	This is APA's basic search engine that includes links to program websites. More detailed information is available at https://gradstudy. apa.org (there is a small fee associated with this service).
https://tinyurl.com/ y5v4gks8	This is a Google map of APA-accredited graduate programs in clinical psychology as of September 2020 (note that it might not be exhaustive).
https://natmatch.com/ psychint/directory/ schools.html	This is a list of doctoral programs that are eligible to participate in the Association of Psychology Postdoctoral and Internship Centers (APPIC) match program. You should only consider programs on this list.

maintains an updated list of all APA-accredited clinical PhD programs. Finally, there are also several websites with search engines. See Table 5.5 for a few of our favorites.

While researching programs, we recommend that you take note of several important parameters that will be included in the program materials or are available upon request. These include the graduation (or attrition) rate, licensure examination pass rate, average number of years to complete the program, percentage of students who obtain APA-accredited internships, and mean GPA and GRE scores. Some programs also include a list of settings in which past students are currently employed and how many of them are licensed, board-certified, etc.

Required Materials

In addition to undergraduate transcripts, programs typically require GRE scores, some type of essay (typically called a "personal statement"), a CV, and letters of recommendation. We will offer a few pointers about each of these and suggest that you read more about them in the *Insider's Guide*.

GRE Although flashbacks to the SAT might tempt you to avoid another standardized test at all costs, the GRE General Test is one of those exams that is simply mandatory, given that ~90% of programs require it (Pagano et al. 2010).[43] It is a computer-based examination that consists of three scales: Verbal Reasoning,

---

[43] It has been suggested that the GRE may not be mandatory for all clinical psychology graduate programs in the future. We suggest that you look into each of your prospective programs to determine whether or not the test is required.

Quantitative Reasoning, and Analytic Writing. The verbal and quantitative scales are more heavily weighted by most programs than is the writing portion. We will not provide much more detail about the test here because you can read all about it directly from the source, at www.ets.org/gre (Educational Testing Service; ETS). That site also contains free test preparation materials and other resources (e.g., search for "Powerprep Practice Tests"). In addition to those resources, some students choose to purchase study books or take formal preparation courses. Keep in mind that you can repeat the exam up to five times in a given year, but it is still in your best interest to earn the highest possible score on your first attempt. Additionally, programs list their average scores in their online materials, so you have an idea as to what score to shoot for in order to gain acceptance into your preferred programs.

In addition to the GRE General Test, some programs require or recommend that applicants take the GRE Psychology Subject Test. This exam assesses the "core of knowledge most commonly encountered in courses offered at the undergraduate level within the broadly defined field of psychology" (https://www.ets.org/). Recommended study strategies for this test include reviewing introductory psychology textbooks and study guides, as well as taking practice exams. You can find out more about the exam and download a practice book at http://www.ets.org/gre/subject/about/content/psychology.

Personal Statement Most programs will require that you submit some form of essay in response to a prompt. A common example is as follows:

> *Summarize your research and career goals, prior educational and work experience, and the reason why you want to become a clinical psychologist. Indicate which faculty member(s) you are interested in working with and describe why you are a good fit with our program. Also, describe your strengths, weaknesses, and any personal life experiences that have shaped your current skills and interests. Two to three double-spaced pages is sufficient.*

Because you will be applying to programs with a strong neuropsychological focus, you will want to showcase relevant research and clinical experience, as well as your passion for the field. Be specific about why/how that program will provide the training you need. Name the neuropsychologist(s) in the program whose work you admire and reference unique aspects of the program such as the excellent neuropsychology coursework and training clinics available to students. Overall, we strongly encourage you to take the time and energy required to weave a concise, yet compelling, well-written story that highlights your strengths and fit with the program. That means beginning the essay-drafting process well before the application deadline (see Table 5.4) and working through several drafts. It will typically be helpful to have friends and family look over your essays before you send them to at least one professor for review.

CV If you do not yet have a CV, you will need to create one for yourself and include it with your application. Most universities offer CV design/editing services and we recommend that you take advantage of this resource (ask your librarian or career center). We have included example CVs on our site, www.NavNeuro.com/book, and you are welcome to use ours as templates. In terms of organization, important sections (in no particular order) include: contact information, education, scientific publications, conference presentations, research experience, clinical experience, scholarships/awards, membership in neuropsychological organizations, teaching experience, and professional references. It is understandable if you do not yet have experiences/accomplishments to add to each of these sections – for example, do not include a *peer-reviewed publications* section if you do not yet have any publications. It is important that this document is formatted uniformly, with experiences consistently presented in either chronological or reverse-chronological order. Keep descriptions concise, relevant, and honest – for example, do not state that you worked at a site for a year when really it was only 6 months. Also, avoid "fluff" or "padding," meaning that you do not artificially lengthen your CV with formatting and/or irrelevant information. Do not include a picture of yourself or interesting facts (the reader does not need to know that you have a poodle named "Oodle"). Finally, ask at least one trusted professor to review your CV prior to submitting it.

Letters of Recommendation If you have followed the advice we laid out earlier in the chapter, then you will have cultivated close relationships with several mentors. Most programs require three letters of recommendation, so choose the three faculty members who know you best and who can speak to your experience and knowledge, as well as your character. For example, we recommend asking someone who is familiar with you as a scientist (e.g., a research advisor) and someone who is familiar with you as a clinician (e.g., a supervisor from one of your clinical internships/ practica). The third letter writer can be another researcher or clinician, or it can be another professor who knows you well. If possible, we suggest that you only consider asking people with doctoral degrees. Finally, it would be a plus if one of the letters is written by a neuropsychologist.

Ask letter writers if they can provide a "strong" letter of recommendation. If they show any signs of apprehension about this, then we advise that you politely retract your request and ask someone else; a neutral letter (e.g., one that is generic and reveals that the writer does not know you well) will not help you and a poor letter will hurt you. We recommend waiving your right to read the letters because this signals to programs that the writers were candid. Make the process of writing the letter as easy as possible by providing letter writers with addressed envelopes and postage, if necessary. We also advise providing your letter writers with a printout of the programs to which you are applying, in addition to deadlines and submission instructions (list it all on one well-organized sheet).[44] Some students also provide their CV and a summary of the activities they completed under the letter writer's supervision, which we believe is a great idea. Ask the writers if they need any additional information. It is incredibly important that your letter writers do not feel rushed, so ask them at least 2 months in advance of your first application deadline.

---

[44] If you have not yet finalized your application list, tell the letter writers you will get it to them soon.

Interview

The vast majority of APA-accredited clinical psychology PhD programs require an interview, either in person or via telephone/videoconference.[45] If a program extends you an offer to interview, it means that you are in their top pool of applicants and they are seriously considering offering you a position in their program. Once you have taken a day or two off (which you deserve!), then it is time to begin preparing. We provide an overview of the interview process in NavNeuro episode 8 (www. NavNeuro.com/08), so we encourage you to listen to that while you are in the preparation phase. Although we created the podcast episode for those applying to a doctoral internship (the final year of graduate school) and postdoctoral fellowship, we believe that a great deal of the information is relevant to the graduate school interview process. To provide more help, we will boil down much of our advice into several bullet points. First, we will cover general interviewing advice and then we will narrow the focus and provide advice specific to graduate school interviews:

General interviewing advice:

• Be prepared.
  - Read about each program; focus on the training requirements, course offerings, faculty clinical and research interests, and practica.
  - Practice interviewing! This includes preparing for common questions (e.g., "what are your strengths and weaknesses?") and role-playing the process with a friend/colleague.
  - Prepare several questions for each program/interviewer.
  - Bring copies of your CV, abstracts of articles that you have authored/coauthored, and printouts of posters that you have presented.
• If you have multiple interviews in a short time span…
  - Plan and book your flights well in advance; build in a safety margin (time between interviews).
  - Mentally prepare yourself for the fact that flights may be canceled or delayed and that this is out of your control.
  - Prioritize safety (e.g., if driving to an interview in a snowstorm).
  - Prioritize sleep.
  - Do your best to cultivate a positive attitude. Have fun and enjoy the journey.
• On the day of the interview…
  - Dress formally.[46]
  - Arrive early.
  - Introduce yourself to other applicants and start small talk. This helps break the tension and allows you to practice talking about yourself and asking questions.

---

[45] Interviewing via videoconference is more common in the COVID-19 era.

- Do your best to stay in the moment. Most people feel nervous in this situation, so you are not alone. Focus on the person in front of you and let your preparation carry you.
- Be positive, enthusiastic, and eager to learn.
- Be aware that you are always potentially being assessed, even when speaking with students or having lunch with other candidates. Always be polite and courteous.
- Pay attention to how you feel about the program (e.g., Are the professors friendly? Do the students seem happy? Does it seem like a collaborative environment?); you are interviewing the program just as much as they are interviewing you.
- After the interview…
  - Within a day or two, send a brief "thank you" email to your interviewers and reiterate your enthusiasm about the program (unless they ask you not to).
  - Immediately write down your impression of the program, perhaps organizing it into a pros and cons list.

Advice specific to graduate school interviews[47]:

- Prepare a compelling, personalized reason why you want to pursue a career in neuropsychology, in place of, "I want to help people," or other common clichés.
- Prepare specific reasons why you chose to apply to each program (e.g., clinical experiences, researchers, available projects, mission statement, student outcomes).
- Review your own experiences and skills that are relevant to psychology and neuropsychology. Be prepared to discuss your background and answer questions about your professional accomplishments.
- If you have an atypical academic or occupational background, find a way to spin your seemingly unrelated experiences so that they are relevant to this new path you have chosen.
- Highlight the clinical applicability of your undergraduate research and how it has prepared you for research at the graduate school level.
- Develop a general idea about the research area you would like to pursue in graduate school and the professors whose interests overlap with your own.

Making a Decision

Hopefully you are in the difficult position of being forced to choose between multiple high-quality programs that have each extended you an offer. If you did your homework up front and carefully selected programs during the application phase, then any one of them would be a solid place to complete your training. At this point, we recommend that you converse with your families/partner, seek out the perspectives of trusted professors, and refer back to each program's outcome data. Update your program rating sheet (see the *Insider's Guide*) and incorporate any new information you learned about each training site during your interview. For example, did you have a good conversation with your prospective advisor? Did you speak to cur-

---

[46] We recommend that you carry your interviewing outfit onto the plane if you fly (to avoid the possibility that the airline loses your luggage).

[47] These also apply to your personal statement.

rent graduate students and did they appear to be genuinely happy with the program? Is the city a good fit for your style and preferences? These are all useful pieces of information that will help you rank order the remaining programs. Arrive at a decision and don't look back.

What If I Didn't Receive Any Offers?

If you did not receive any offers, then take a moment to reflect on what might have been the issue with your application and/or choice of programs. First of all, recall that graduate programs in clinical psychology are competitive. You may not have done anything wrong and it is often impossible to know precisely why offers were not extended. We recommend that you take time off for a few days and do something fun to distract yourself, and then dust yourself off and get to work at improving any aspects of your application that you feel may have been lacking. This may mean that you retake the GRE after completing a formal preparation course, that you retake a college course or two to replace that "C" with an "A," that you take on another research project that leads to a peer-reviewed publication, and/or that you seek out more clinical experience. Discuss your situation and options with trusted professors and advisors. Some students postpone graduation so they can complete some of these activities at their university, while others seek out opportunities after graduation (i.e., as a post-baccalaureate). You might even consider applying to a master's program in psychology; this would allow you to verify that a doctorate is the right choice for you, and it would provide you with additional knowledge and experience.[48] As we noted earlier, many people do not go straight into a doctoral program after finishing college (Zimak et al. 2011). Additionally, as a potential silver lining, the same study found that students who postponed graduate school after college reported fewer thoughts of withdrawing from the program relative to their counterparts who went straight from undergraduate to graduate study, possibly because the extra time increased their confidence about pursuing this path (Zimak et al. 2011).

****

Regardless of the outcome, do not forget to thank everyone who helped you through the application process (e.g., letter writers and those who reviewed CVs, personal statements, etc.). Also, take time to decompress after this long ordeal. Finishing college and completing applications are major tasks, and you deserve some time to yourself.

---

[48] However, keep in mind that the vast majority of clinical psychology PhD programs do not require a master's degree, and most students do not enter with one (Norcross et al. 2010). In fact, one study found that most PhD programs were equally likely to accept a baccalaureate-level student as a student with a master's degree (Littleford et al. 2018).

Action Steps:
Follow the application preparation timeline laid out in Table 5.4.
Look for APA-accredited PhD programs in clinical psychology with strong neuropsychological training.
Organize a spreadsheet with programs of interest, including relevant data on each program (see Table 5.5).
Apply to ~12 programs, attend interviews, and select the program that is the best fit for you

Conclusion

We wish you the best of luck with the remainder of your time in college and the graduate school application process. As we mentioned at the outset of the chapter, we have provided you with a large number of specific recommendations and we have set the bar high. However, we don't want your grades or research to suffer because you are attempting to follow each and every recommendation "to a T." If you find yourself becoming overwhelmed, focus on the essentials. To reiterate, those are as follows: (1) learn about psychology and neuroscience, (2) do well in your courses, (3) build relationships with professors, (4) become involved in research, (5) consider the aspects of neuropsychology that interest you the most, (6) prepare for the graduate school application process well in advance, and (7) have fun at every step of the way. If you're able to manage these, then the other details will be icing on the cake.

We'll see you in the next chapter, after you have been accepted to a graduate program!

References

American Psychological Association. (2017). *Ethical principles of psychologists and code of conduct* (2002, amended effective June 1, 2010, and January 1, 2017). https://www.apa.org/ethics/code/

American Psychological Association. (2019). *The publication manual of the American Psychological Association* (7th ed.). American Psychological Association.

Baker, D. B., & Benjamin, L. T. (2000). The affirmation of the scientist-practitioner: A look back at Boulder. *American Psychologist, 55*, 241–247.

Doran, J. M., Kraha, A., Marks, L. R., Ameen, E. J., & El-Ghoroury, N. H. (2016). Graduate debt in psychology: A quantitative analysis. *Training and Education in Professional Psychology, 10*(1), 3–13.

Dunn, D. S. (2011). *A short guide to writing about psychology* (3rd ed.). Pearson Education, Inc.

Driskell, L. D., Del Bene, V. A., & Sperling, S. A. (2020). How to become a competitive neuropsychology intern and postdoc applicant. *Lecture*. https://knowneuropsych.org/how-to-be-a-competitive-fellow-intern/

Hannay, H. J., Bieliauskas, L. A., Crosson, B. A., Hammeke, T. A., Hamsher, K. deS., & Koffler, S. P. (1998). Proceedings of the Houston conference on specialty education and training in clinical neuropsychology. *Archives of Clinical Neuropsychology, 13*(2), 157–158.

Huss, M. T., Randall, B. A., Patry, M., Davis, S. F., & Hansen, D. J. (2002). Factors influencing self-rated preparedness for graduate school: A survey of graduate students. *Teaching of Psychology, 29*(4), 275–281.

Littleford, L. N., Buxton, K., Bucher, M. A., Simon-Dack, S. L., & Yang, K. L. (2018). Psychology doctoral program admissions: What master's and undergraduate-level students need to know. *Teaching of Psychology, 45*(1), 75–83.

Nicol, A. A. M., & Pexman, P. M. (2010a). *Presenting your findings: A practical guide for creating tables* (6th ed.). American Psychological Association.

Nicol, A. A. M., & Pexman, P. M. (2010b). *Displaying your findings: A practical guide for creating figures, posters, and presentations* (6th ed.). American Psychological Association.

Association of Psychology Postdoctoral and Internship Centers (APPIC). (2019). *APPIC match: 2011–2019: Match rates by doctoral program*. https://www.appic.org/Portals/0/downloads/APPIC_Match_Rates_2011-2019_by_UniversityV2.pdf

Norcross, J. C., Ellis, J. L., & Sayette, M. A. (2010). Getting in and getting money: A comparative analysis of admission standards, acceptance rates, and financial assistance across the research–practice continuum in clinical psychology programs. *Training and Education in Professional Psychology, 4*(2), 99–104.

Norcross, J. C., Castle, P. H., Sayette, M. A., & Mayne, T. J. (2004). The PsyD: Heterogeneity in practitioner training. *Professional Psychology: Research and Practice, 35*(4), 412–419.

Norcross, J. C., Sayette, M. A., & Pomerantz, A. M. (2017). Doctoral training in clinical psychology across 23 years: Continuity and change. *Journal of Clinical Psychology, 74*(3), 385–397.

Norcross, J. C., Sayette, M. A., Stratigis, K. Y., & Zimmerman, B. E. (2014). Of course: Prerequisite courses for admission into APA-accredited clinical and counseling psychology programs. *Teaching of Psychology, 41*(4), 360–364.

Pagano, V., Wicherski, M., & Kohout, J. (2010). *2008–09: Test scores and requirements for Master's and doctoral students in U.S. and Canadian graduate departments of psychology*. American Psychological Association. https://www.apa.org/workforce/publications/10-grad-study/test.

Petrella, J. K., & Jung, A. P. (2008). Undergraduate research: Importance, benefits, and challenges. *International Journal of Exercise Science, 1*(3), 91–95.

Powell, R. A., Honey, L. P., & Symbaluk, D. G. (2016). *Introduction to learning and behavior* (5th ed.). Cengage Learning.

Rabin, L. A., Paolillo, E., & Barr, W. B. (2016). Stability in test-usage practices of clinical neuropsychologists in the United States and Canada over a 10-year period: A follow-up survey of INS and NAN members. *Archives of Clinical Neuropsychology, 31*(3), 206–230.

Ready, R. E., & Santorelli, G. D. (2014). Values and goals in clinical psychology training programs: Are practice and science at odds? *Professional Psychology: Research and Practice, 45*(2), 99–103.

Ritchie, D., Odland, A. P., Ritchie, A. S., & Mittenberg, W. (2012). Selection criteria for internships in clinical neuropsychology. *The Clinical Neuropsychologist, 26*(8), 1245–1254.

Sayette, M. A., & Norcross, J. C. (2020). *Insider's guide to graduate programs in clinical and counseling psychology: 2020/2021 edition*. The Guilford Press.

Schaffer, J. B., Rodolfa, E., Owen, J., Lipkins, R., Webb, C., & Horn, J. (2012). The examination for professional practice in psychology: New data–practical implications. *Training and Education in Professional Psychology, 6*(1), 1–7.

Schulte, B. A. (2003). Scientific writing & the scientific method: Parallel "hourglass" structure in form & content. *The American Biology Teacher, 65*(8), 591–594.

Smith, G., & CNS. (2019). Education and training in clinical neuropsychology: Recent developments and documents from the clinical neuropsychology synarchy. *Archives of Clinical Neuropsychology, 34*(3), 418–431.

Sweet, J. J., Klipfel, K. M., Nelson, N. W., & Moberg, P. J. (2020). Professional practices, beliefs, and incomes of postdoctoral trainees: The AACN, NAN, SCN 2020 practice and 'Salary Survey'. Archives of Clinical Neuropsychology.

Whiteside, D. M., Guidotti Breting, L. M., Butts, A. M., Hahn-Ketter, A. E., Osborn, K., Towns, S. J., et al. (2016). 2015 American Academy of Clinical Neuropsychology (AACN) student affairs committee survey of neuropsychology trainees. *The Clinical Neuropsychologist, 30*(5), 664–694.

Zakaria, F. (2015). *In defense of a liberal education.* W. W. Norton & Company.

Zimak, E. H., Edwards, K. M., Johnson, S. M., & Suhr, J. (2011). Now or later? An empirical investigation of when and why students apply to clinical psychology PhD programs. *Teaching of Psychology, 38*(2), 118–121.

Chapter 6

Doctoral Training

> Be patient. It takes a long time to get a PhD with a focus in clinical neuropsychology and sometimes it feels never ending. You complete your doctorate, you finish a clinical internship, you go through the two year postdoc… it just feels like you're never going to finish. It does take a long time to get to that point, but once you're independent, there are so many ways to keep your career meaningful and exciting for years to come.
>
> – Peter Arnett, PhD

Now that you are in graduate school (congrats, by the way!), the real work of honing your psychological and neuropsychological skills is about to begin. The next several years are likely to be the most academically challenging of your life. You will be balancing courses, research, clinical work, and professional development at an unprecedented level. As such, it may be tempting to view graduate school as a series of hoops to jump through. However, we encourage you to approach this time in your life with a growth mindset, meaning that you face challenges head-on because you understand that your abilities are malleable and you are seeking to cultivate your knowledge and skills (Dweck 2007). Graduate school is your opportunity to begin building expertise in an area of great interest to you. Your hard work at this stage will pay off in dividends down the road because the experience you gain lays the foundation for the rest of your career.

Successfully traversing graduate school requires many skills. Principal among them are perseverance, the ability to delay gratification, self-discipline, organizational proficiency, flexibility, interpersonal skills, and a desire to grow and learn. The fact that you have completed college and gained acceptance into a doctoral program means that you already possess these abilities to some degree, and they

J. A. Bellone, R. Van Patten, *Becoming a Neuropsychologist*,
https://doi.org/10.1007/978-3-030-63174-1_6

will be tested repeatedly over the next few years. Additionally, keep in mind that organizational proficiency and conscientiousness can be altered and improved. The field of clinical psychology is built to a large degree on the premise that behavior change is possible; so, if you struggle with disorganization, procrastination, or lack of motivation, don't despair! Use the tips and strategies provided in this book and work toward improving your efficiency and productivity. This will allow you to achieve success in graduate school and have fun while doing so.

Training Guidelines

Before we launch into the roadmap for graduate school, we want to tell you about the guidelines that will dictate your training in clinical neuropsychology from here on out. These guidelines are a big deal in our field, and are drawn on heavily by the credentialing boards to determine eligibility for board certification. As such, we will revisit them where applicable throughout the rest of the book.

Houston Conference Guidelines In 1997, prominent members of the profession held a conference in order to determine "a model of integrated education and training in the specialty of clinical neuropsychology" (Hannay et al. 1998; p. 1). The resulting policy statement, commonly referred to as the "Houston Conference guidelines," continues to act as the primary set of recommendations for neuropsychological training in graduate school, internship, and postdoctoral fellowship, with wide adoption among programs (Sweet et al. 2012).[1] Although technically aspirational (as opposed to mandatory), adhering to these recommendations is a sure-fire way to remain eligible for board certification in clinical neuropsychology, which solidifies your standing as a competent, skilled practitioner in the field.[2] As such, we suggest that you put this book down now (don't worry, it's only for a few minutes!) and read the Houston Conference (HC) guidelines.[3] It won't take long and it will give you a broad overview of the path forward, as well as an understanding of the specific knowledge and skills needed to become a clinical neuropsychologist.

* * * *

As you saw in the document, it is important for doctoral students to receive specialized neuropsychological instruction. The HC guidelines document states, "The foundation of brain-behavior relationships should be developed to a considerable

---

[1] An important document regarding training in clinical neuropsychology that predates the HC guidelines is the International Neuropsychological Society-Division 40 report (1987). To read more about the history leading up to the HC, see Bieliauskas and Steinberg (2005), and Bodin et al. (2016).

[2] As of the time of this writing, board certification is a clinical credential in neuropsychology and is far less useful for researchers. Consequently, if your career will be entirely in the research domain, board certification is not necessary.

[3] You can find the document here: www.NavNeuro.com/HCG.

degree at this level of training" (Hannay et al. 1998; p. 3). Of note, the authors acknowledge that programs will vary in the degree to which neuropsychology is emphasized, and that students should also receive training in "generic psychology and clinical core" (p. 3). As such, we encourage you not to worry that your program is not "neuropsych enough," especially in the first few years. Generalist training in clinical psychology is the backbone upon which specialty knowledge in neuropsychology is built. The more time that passes, and the further you proceed in your training, the more you will be able to focus and home in on neuropsychology.[4] Indeed, when you begin completing your advanced training, specialty neuropsychological instruction will predominate and you will acquire the depth of knowledge that provides neuropsychologists with the brain/behavior tools to transition into their career as a skilled practitioner.[5]

These are the Houston Conference guidelines, not commandments.

– Anthony Stringer, PhD, ABPP-CN

The hope would be that everyone would stay on the yellow brick road, with organized training and didactics at each stage of the process. But it is important to know that you can reach the same end stage with different experiences at different steps. I think that's what was really intended by the Houston Conference.

– Peter Dodzik, PsyD, ABPdN, ABN

Taxonomy for Education and Training in Clinical Neuropsychology[6] More recently, a taxonomy has been created by the Clinical Neuropsychology Synarchy (CNS).[7] As noted by Sperling et al. (2017), the Taxonomy "was developed to be consistent with and supportive of the HCG [Houston Conference guidelines], not to supplant them, by offering specific and common definitions for programs to apply to their HCG-consistent training" (p. 819). Four degrees of intensity of neuropsychological training (Major Area of Study, Emphasis, Experience, Exposure) are listed at each of the four training stages (doctoral, internship, postdoctoral, and post-licensure). This includes minimum "requirements" to meet neuropsychological criteria for coursework, clinical training, didactics, and research.

At the time of this writing, neuropsychology boards have not mandated that students receive training consistent with any specific intensity level at any stage of training, with the exception of postdoctoral fellowship, for which the boards require the equivalent of a "Major Area of Study." Furthermore, the language of the Taxonomy has not been uniformly adopted yet, so it cannot be relied upon as the

---

[4] If you are already in a clinical psychology graduate program that does not offer any neuropsychological training, see the *Applying to practicum sites* section for advice on securing neuropsychological experience.

[5] As we were writing the book, there was a proposal for holding a second conference to update the HC guidelines.

[6] Referred to as "Taxonomy" from here on out.

[7] For access to the full Taxonomy and to see requirements for other psychological specialties, see https://www.cospp.org/education-and-training-taxonomies.

only source of information to determine the quality of a program's neuropsychological training. However, look out for these terms (e.g., "our program offers a 'Major Area of Study' in clinical neuropsychology") in the near future because it is likely that more programs will adopt them. We are in full support of implementing this language, as we believe that standardized nomenclature and practices would drastically improve clarity for prospective students researching pre- and postdoctoral training programs. As such, we will list the criteria for "Major Area of Study" at each stage and recommend that you do your best to align your training with these criteria. For more information pertaining to the Taxonomy, see Smith and CNS (2019). Of note, this article also lists entry-level competencies for clinical neuropsychologists.

Overview and Timeline

There are many steps to completing a doctoral degree in clinical psychology, including coursework, research involvement, clinical experience, a comprehensive exam, and a formal clinical internship. We will provide advice and guidance pertaining to each of these tasks throughout this chapter. For now, Table 6.1 shows a timeline for a typical PhD program.[8]

Although some students graduate in less than 6 years, we suggest that most people stick to their program's standard timeline rather than trying to shave off a year from their training. The reason for this is that there is a great deal to learn and many experiences to be had, and every decision comes with an opportunity cost. Consequently, you need time in order to soak it all in. For example, entering the internship year early allows for fewer clinical practica and scientific publications in graduate school. In order to ensure that you acquire all of the necessary skills and "get your money's worth" from graduate school, it is important that you take your time and complete each and every task at a high level. This will also make you a competitive applicant for internships, fellowships, and, ultimately, jobs in neuropsychology. Of course, some people may seek to expedite the graduate school process for personal and/or financial reasons, and so, under certain circumstances, it can be the right decision to earn the degree in five years rather than six. The most important idea here is getting high-quality training.

One additional note before we move on. In most doctoral programs, each student obtains a master's degree after completing the thesis and other requirements, typically at the end of the second or third year. This occurs automatically in the course of the training; students are not required to apply to separate master's and doctoral programs. While the degree will be dropped from your credentials once you obtain a doctorate (you will be, e.g., "Keshawn Williams, PhD"), the master's degree provides more clout during the latter half of graduate school, and this can open doors

---

[8] Much of this chapter will assume you are in a clinical psychology PhD program that adheres to a scientist-practitioner (Boulder) model. However, the vast majority of the information will also be applicable to PsyD students and clinical scientist PhD students.

Table 6.1 Example of a PhD program timeline

Year	Tasks
1	Complete coursework.
	Contribute to research projects under your advisor's guidance.
	Generate research questions, aims, and hypotheses for the master's thesis; begin the writing process.
	Identify clinical opportunities that fit with your career goals.
2	Complete coursework.
	Contribute to research projects under your advisor's guidance.
	Propose and carry out your thesis project.
	Apply for a practicum[a] and begin acquiring clinical hours.
3	Complete coursework.
	Contribute to research projects under your advisor's guidance.
	Defend your master's thesis.
	Generate research questions, aims, and hypotheses for the doctoral dissertation; begin the writing process.
	Build clinical experience through formal practica.
4	Finish coursework.
	Contribute to research projects under your advisor's guidance.
	Propose and begin data collection for your dissertation project.
	Complete additional clinical practica.
	Pass the comprehensive examination.
5	Apply to internship.
	Defend your dissertation prior to beginning internship (if possible).
	Complete additional clinical practica.
6	Complete internship.
	Apply to postdoctoral programs.

Note: Not included here are professional development activities such as attending conferences and workshops, serving on committees, etc.
[a]This goes by other names including "externship" and "placement"

for you. For example, if you apply to teach a course, complete an advanced clinical practicum, or serve on a professional committee, a master's degree will be looked upon more favorably than a bachelor's degree.

Now, let's discuss each of the major aspects of graduate school in more detail.

Coursework

Programs differ in the types of classes offered and in the number of units required to graduate. Typically, students have a full course load during the first year, with a gradual tapering down over time as other responsibilities ramp up. Prior to the beginning of your first semester, we recommend asking the program administrator for a list of classes projected to be offered over the next few years and also consulting the graduate school handbook for a sample schedule. Next, you can sit down with your advisor and talk through a rough timeline in terms of when you plan to take each class. This approach will help you avoid accidentally missing a graduate school requirement and will allow you to answer questions such as, "Which electives should I take?" and "When should I register for one class versus another?"

You will be required to take both core and elective courses. Although many of the core classes will not appear to be directly pertinent to neuropsychology, the knowledge you gain will make you a well-rounded psychologist, which is critical in our field. That being said, it is also important to begin acquiring detailed specialty knowledge that will be relevant to your particular career path.[9] In Table 6.2, we provide examples of courses offered by most clinical psychology programs with neuropsychology concentrations. Keep in mind that it is not required for all of your electives to be neuropsychology specific. Depending on your interests, you might benefit from a specialty course in structural equation modeling, family therapy, or health psychology, for example.[10] If there is an interesting neuropsychology course that is not offered by your program, you have a few options. First, you can ask your advisor if it is possible for you to take a "reading course" – this is a specialty elective that is not part of the traditional curriculum, but that is made available when requested by one or a few students. If a neuropsychology professor agrees to teach a reading course, the department will then formalize it such that it counts for credits on the student's transcript. Alternatively, if a reading course is not available, you can look to neuropsychological organizations for help. We recommend the National Academy of Neuropsychology for this – go to https://www.nanonline.org/ → Continuing Education → DistanCE E-Learning for more details.

Table 6.2 Example core and elective courses

Core	Neuropsychology-focused elective
Advanced statistics	Fundamentals of human neuropsychology
Research methods	Neuropsychological assessment
Psychopathology	Neuroscience
Clinical interventions	Neuroanatomy
Clinical assessment	Neurodegenerative diseases
Developmental psychology	Psychometric foundations
Social psychology	Behavioral neurology
Ethics and professional issues	Psychopharmacology
Human diversity	
History of psychology	
Cognitive psychology	

---

[9] Per the Taxonomy, a student must take a minimum of three neuropsychology courses (i.e., "prominently address areas outlined in the Houston Conference Guidelines") for neuropsychology to be considered a Major Area of Study. Keep in mind, however, that a Major Area of Study in neuropsychology is not currently required for board certification.

[10] We suggest that you file away all syllabi and document any learning experiences relevant to neuropsychology because this will come in handy when applying for board certification.

Graduate school courses are similar to college courses in many ways, but there are some important differences:

1. The class size will typically be smaller; this allows for more active participation and less pure lecture.[11]
2. Assignments are often essays and presentations, with fewer exams (especially multiple-choice tests).
3. Professors typically present content at a higher level and expect a higher quality product (e.g., essays, presentations).
4. It is expected that students will earn As and Bs in general courses, and As in specialty courses (for you, those are the neuropsychology courses). Anything lower is typically considered sub-par.

We recommend always putting forth your very best effort in all of your classes, but keep in mind that grades are not emphasized as strongly in graduate school as they were during your undergraduate years. More important than the letter grade is the knowledge and skills you gain through your coursework. Additionally, you will be juggling other very important tasks such as research and clinical work, and these areas are heavily emphasized by internship and fellowship programs, so we recommend that they take priority and that you do not concern yourself with the difference between a 95 and a 97 on an exam.

Action steps:

 Ask for a list of available courses. Map out your coursework over the next few years.

 Take all required courses and as many neuropsychology-related electives as possible.

 Focus on learning the material more than the grades.

 Eliminate distractions during class time.

Research Involvement

General Research

One of the great things about our field is that we're constantly striving to draw conclusions on as rigorous and on as empirical a basis as possible. – Dean Delis, PhD, ABPP-CN

---

[11] There are multiple benefits to active educational practices (asking questions, summarizing, discussing) compared to simply listening passively to lectures (Gregory et al. 2006; Magana et al. 2017; Michel et al. 2009).

One of the primary reasons for attending graduate school (especially a PhD program) in psychology is to receive a strong scientific education. The research training afforded by clinical psychology graduate programs sets the field apart from other clinical specialties, which often do not provide trainees with the same level of rigorous instruction as we receive in psychology. The clinical psychology graduate school research experience also differs from that which is provided in college in several ways: (1) you will be expected to take on a heavier workload, (2) you will be carrying out more complicated tasks, and (3) you will begin learning how to pioneer original, independent scientific work. Overall, this is your chance to experience what it is like to be fully immersed in the investigative process and to pursue research questions that are of great interest and importance to you.

The one external factor that will impact your research experience and overall success/enjoyment in graduate school more than anything else is the skillset and interpersonal style of your advisor. This is the single person with whom you will spend more time than anyone during your five years of pre-internship graduate study. Your schedule with them will likely include weekly individual supervision, as well as group-based lab meetings. Your thesis and dissertation will be conducted in their lab and they will mentor and advise you throughout these critical undertakings. You will seek their guidance regarding which classes to take, which clinical practica to pursue, and which internship programs to apply to. They will likely teach several of your courses and they will be your advocate throughout this crucial period of time in your professional development. They will also write you letters of recommendation for clinical internship, postdoctoral fellowship, and grants and awards. Clearly, this is an incredibly important person; you might even think of them as your academic parent. So, hopefully you heeded our advice in the previous chapter and put a great deal of weight on the productivity, reputation, and personalities of the potential faculty advisors at your program.

Programs differ in terms of how a research advisor is selected. In some programs, the applicant is matched to an advisor upon admission, while other programs first accept students and then allow them time to choose the faculty mentor. Obviously, it will be in your best interest to work with a neuropsychologist, or at least someone who has knowledge about the field and who specializes in neuroscience, cognitive psychology, or health psychology.

Importantly, some programs allow students to change their advisors if the relationship is not working out. This is not common, because people typically identify a good mentor on the first try. But if you find yourself in a toxic situation, you may have options. Read your program's Student Manual and/or Program Handbook and learn about the procedures for switching mentors. Often, you will talk to your current advisor, as well as the Director of Clinical Training (DCT), about what is not working in the relationship and about what you hope to improve through an alternative match. If they agree that it is the right decision, you will move forward. This is not something to be taken lightly, but it can be the right decision for some people. You cannot always predict how a relationship will evolve, and it may be well worth the effort and initial discomfort to make a switch in your first few years if it ultimately leads to a healthy and productive professional relationship for the rest of graduate school and beyond.[12]

* * * *

Once you have selected an advisor and are settled into their lab, we suggest that you begin working on research projects as soon as possible. The priorities here are (a) enhancing your knowledge of statistics, research methods, and the scientific literature in your area of study, (b) acquiring peer-reviewed publications, (c) writing grants, (d) presenting posters and papers at conferences, and (e) serving as a peer reviewer for scientific journals[13]. As in all aspects of graduate school, the acquisition of new knowledge and skills is paramount. From there, as we discussed in Ch. 5, publications are the best way to advance your career; this is especially true if you want to pursue a position at an academic institution, but also important if you are interested in a purely clinical career.[14] Each and every scientific paper is a tall task, requiring a great deal of effort and patience. So, start early and be prepared. Discuss available projects with your advisor, identify those that appear to be most interesting and relevant to you, and then offer your services. If necessary, you can ask your advisor to help you set goals and create deadlines for various aspects of the process (e.g., analyzing data, writing). Some people find that this added structure improves their productivity.

In terms of which research projects to work on, there are two general strategies to consider, with one favoring breadth and the other prioritizing depth. We both adopted a "shotgun" approach, where we became involved in a wide variety of projects in different research areas within psychology. This allowed us exposure to a range of different research topics (e.g., animal models of various neuropathologies, neuropsychological effects of nutritional changes, the conceptualization and treatment of gambling disorder), as well as experience working with a large number of different supervisors and collaborators. For us, prioritizing breadth was also the best way to maximize the number of publications and formal presentations. In part, this was because we were not provided with enough opportunities in a particular niche area of study and, in part, it was because we are interested in a wide array of topics.

The alternative to the shotgun approach is a laser-focus in one area. This builds great depth of knowledge and experience in a particular scientific domain and can be helpful if two criteria are met: (a) the student is very confident about their interest in one research topic, and (b) the training program offers plenty of research opportunities in that area (typically, in the research advisor's lab). An advantage of the

---

[12] We strongly suggest that no matter how upset you are with your research advisor, you conduct yourself professionally. The neuropsychology community is quite small, and it is helpful to avoid burning bridges if possible.

[13] See Duff et al. (2009).

[14] It is impossible for us to provide a firm recommendation in terms of a goal for the number of publications in graduate school. There are simply too many moderating factors (e.g., clinical versus research focus, impact of each paper, area of study). The best advice is, "the more high-quality publications, the better."

laser approach is that it is consistent with most researchers' career goals. The world is far too complex for any single person to be an expert in everything, so scientists specialize and subspecialize. They become known for their work in a small corner of the world, and they build a skyscraper of knowledge in that domain. For example, Dr. Adam Woods is a research neuropsychologist at the University of Florida. He is interested in interventions to improve thinking and emotional functioning as people age. Specifically, he works in noninvasive brain stimulation to sustain and improve cognitive and mental health. Even more specifically, he is renowned for his knowledge in a particular type of noninvasive brain stimulation called transcranial direct current stimulation, or tDCS. See what we mean?

So, shotgunning[15] provides a diversity of experiences and a wide range of knowledge, while the laser approach builds expertise in one particular domain. Of course, there is no right or wrong answer, and there are gradations of breadth and depth between the two extremes. You could focus most of your energy on your area of greatest interest, while dabbling in secondary interests as well. Consider your circumstances and the topics that spark your curiosity, and talk to your advisor about your options.

Thesis and Dissertation

Most programs require the completion of both a thesis and dissertation. We will provide a few tips and pointers here, but a full roadmap through these projects is beyond the scope of our book. For detailed guidance, we suggest that you read *Dissertations and Theses From Start to Finish: Psychology and Related Fields* (2019) by Bell, Foster, and Cone.

First and foremost, the thesis and dissertation are educational research projects designed to teach students how to conduct rigorous and relevant scientific work. We cannot overemphasize the importance of this foundational training. With the guidance of your advisor, you will complete an entire study from start to finish. This includes learning and understanding the scientific literature in a particular area, generating relevant research questions to fill in gaps in the literature, crafting aims and hypotheses to answer the research questions, proposing and justifying the entire study, collecting and analyzing the data, writing up the results in a full manuscript, and defending your work against the scrutiny of others. If that does not sound like a monumental task, it will feel like one once you get started. But you will have your advisor, as well as institutional resources and support on your side, so take your time and enjoy the ride.

---

[15] Not to be confused with a certain alcohol consumption technique!

Now, let's get specific with a few recommendations. First, select research questions that are relevant to neuropsychology.[16] Not only will this accelerate your acquisition of knowledge in the field but it is also looked upon favorably by neuropsychology-focused internships and fellowships,[17] so it will help you in your applications later on down the road. Second, if you are interested in pursuing direct patient care in your career, then we recommend that you focus your thesis and dissertation on clinically relevant research questions rather than purely theoretical or "basic" (non-applied) areas. Third, manage your expectations for these projects. Remember, these are training exercises, not Nobel prize worthy scientific revelations. Paradigm-shifting studies often take years, or even decades to complete,[18] and they can cost millions of dollars. You are not at that level yet. Many students try to aim too high and bite off more than they can chew with their thesis and/or dissertation, and then they fall behind their projected timeline as a result.[19] This creates a great deal of stress and it can hold you back in your career, so talk to your advisor about a realistic, manageable project.

> Fortuitously, one of the members of my committee said to me, 'Glenn, we don't expect your dissertation to be the best research you'll do in your career.' And that really mattered to me. It made me recognize that, yeah, this is my first real independent research project, not my last one.
>
> – Glenn Smith, PhD, ABPP-CN

A Note to PsyD Students

Students who are pursuing a PsyD may not be required to complete a thesis and dissertation, or at least not the same format as the research project described above. The capstone PsyD project is sometimes referred to as a "doctoral project," and it is often an extensive literature review on a relevant topic. Now, even if research is not emphasized to the degree that it is in PhD programs, we still recommend that you build up research experience during your graduate study (including publications and oral presentations); this will make you both a more competitive applicant down the line and a better practitioner-scholar.[20]

---

[16] Per the Taxonomy, a student must complete a neuropsychology-related dissertation or research project in order for neuropsychology to be considered a Major Area of Study (Smith and CNS 2019).

[17] Per a recent survey of 88 training directors of postdoctoral fellowships, 50% stated that it is "somewhat important" for a dissertation to have a neuropsychology focus, 39% stated that it is "very important," and 6% stated that it is "essential" (Driskell et al. 2020).

[18] A great example of this is the Framingham Heart Study, which began in 1948 in Framingham, Massachusetts, and is still ongoing. They have collected data from thousands of people across four generations and published over 1,000 peer-reviewed articles based on their data. Much of what we know about heart disease today comes from this gargantuan project.

[19] The common saying, "The best dissertation is a finished dissertation" is relevant here.

[20] See Ch. 5 for an explanation of the differences between PsyD and PhD degrees.

Action steps:

 Select a research advisor who is well-published, student-friendly, reliable, and a good personality fit.

 Begin working on multiple research projects as soon as possible.

 Publish and present your research findings as frequently as possible.

 Complete a manageable and clinically relevant thesis and dissertation that pertain to neuropsychology.

Clinical Training

Introduction

Gaining clinical experience is another hallmark feature of graduate school in our field, as it allows the student to apply the knowledge acquired from research and coursework in the service of patient care (Nelson et al. 2015).[21] Although you may have received some form of clinical training as an undergraduate, your initial graduate practicum will likely be the first time you deliver psychological services one-on-one with patients. Most students find this both exciting and nerve-racking, and this mix of emotions is entirely normal. First, remember that you will be supervised by licensed psychologists throughout the entire process, so take advantage of their wisdom. Second, consider our advice, below.

Most students will complete at least three separate clinical practica prior to internship.[22] These are part-time (e.g., 1–3 days per week)[23] and they vary in length. The Association of State and Provincial Psychology Boards (ASPPB) recommends that at least 50% of practicum time be spent in "service-related activities" (e.g., intervention, assessment, report writing) and at least 25% of the experience involve direct patient contact (ASPPB, 2009, p. 8). As we noted in Table 6.1, PhD students

---

[21] Clinical experience is important even for those of you who are interested in a purely research-focused career in neuropsychology. Specifically, the brain-behavior landscape is inundated with clinical populations, as we have learned a great deal about brain functioning from people with brain injuries and diseases (see Chapter 1). Consequently, neuropsychological researchers reflect back upon their clinical training in graduate school and pull from it to inform both their general understanding of the scientific literature and the specific research questions that they address.

[22] See Sweet et al. 2020a for results from a survey of postdoctoral fellows asked how many neuropsychological and psychotherapy/intervention practica they completed during their doctoral training.

[23] Per Nelson et al. 2015, a practicum typically consists of 16 hours per week across an academic year and should involve a minimum of eight hours per week.

typically complete their first practicum in year two or three and then continue to acquire diverse clinical experiences for the rest of their time in graduate school.

Psychological Interventions

The clinical question that we receive most frequently from graduate students interested in our field is, "How soon should I begin to specialize in neuropsychology?" While there is no definitive answer to this question, we recommend focusing on broad, generalist training early in graduate school rather than seeking to hyperspecialize immediately.[24] As we have mentioned several times, a comprehensive understanding of the field of clinical psychology is essential to competent and skilled neuropsychological practice. Specialty training in neuropsychology will increase as you progress in your graduate training.

> Don't forsake your general training in the interest of acquiring as much neuropsychological training as you can. You will have plenty of time to specialize in your fellowship and beyond. Make sure that you come out of your training well-rounded.
>
> – Neil Pliskin, PhD, ABPP-CN

> Remember that every good clinical neuropsychologist is a clinical psychologist first. The further I go in my career, the more I rely on good interviewing skills and people skills, and I am so convinced that these are some of the most important and powerful things that we can do.
>
> – Robin Peterson, PhD, ABPP-CN

Most programs will have an in-house mental health clinic for the initial practicum (often called the "Psychological Services Center" or something similar), and this is often where the training begins. Although neuropsychology currently centers around assessment, clinical psychology focuses heavily on psychological interventions, so your generalist training will invariably involve delivering treatment to patients. Most programs/supervisors adhere to a cognitive-behavioral therapy (CBT) orientation (Wright et al. 2017), so you will likely receive training in these techniques and focus most of your intervention experience in this modality. Examples of other popular and useful evidence-based therapies to become familiar with include motivational interviewing (Miller and Rollnick 2013), acceptance and commitment therapy (Hayes et al. 2012), solution-focused therapy (O'Connell 2012), and interpersonal psychotherapy (Frank and Levenson 2010).

If available, we have also found training in group-based therapy and couples/family therapy to be very beneficial. For example, the skills you develop by way of this training will help you through challenging feedback sessions when your older patient's family members disagree about how best to deal with his cognitive and

---

[24] If you are set on gaining neuropsychological experience during your first clinical year, we recommend that the practicum either contain a large generalist component (e.g., a setting with both cognitive assessment and psychological intervention) or that you consider completing two practica simultaneously (one in a neuropsychological setting and one in a generalist setting).

functional decline, or when your young patient's parents are unsure about whether or not their child should be placed in special education classes.[25] Another excellent type of clinical experience can be gained by working at a psychiatric facility (either locked/inpatient or outpatient), as you will be exposed to more severe forms of mental illness, including relevant psychological interventions for people with these conditions.

Ideally, you will have the opportunity to shadow your supervisor and/or an upperclassman before seeing patients on your own because observing skilled clinicians is a great initial learning experience. However, even if you have done so, the first several sessions of therapy will likely feel awkward and challenging. A psychotherapy session is like no other interpersonal experience. It is very common to feel out of one's comfort zone, and the imposter syndrome often surfaces in this situation (e.g., "How can *I* help this person through an episode of depression?"). Here are some general tips for your initial foray into one-on-one therapy:

- Communicate frequently with your supervisor; ask questions.
- Consult with your clinical supervisors about arranging mock therapy sessions with classmates.
- Study treatment manuals and other resources (e.g., the Wright et al. 2017 book).
- Watch professional videos of mock therapy sessions and discuss them with colleagues and supervisors.
- Lean on the knowledge you have gained in your coursework (e.g., active listening, open-ended questioning, and summarizing).
- Learn to tolerate silence.
- If you have personal experience in therapy, reflect upon how it felt for you, how the therapist approached your treatment, and how you may have benefited from it.

Neuropsychological Services

Similar to the utility of early generalist training in clinical psychology, we believe that it is important to receive broad neuropsychological training as well.[26] It is sometimes tempting for people to immediately hyperspecialize in the specific niche of neuropsychology in which they are particularly interested (for example, working in a sports concussion clinic where all of the cases are adolescent and young adult athletes who sustained traumatic brain injuries). However, at this stage in your training, we recommend that your goal be to provide services in a variety of settings (e.g., inpatient, outpatient, rehabilitation) and to people of various age cohorts, cultural backgrounds, and clinical syndromes (e.g., neurodevelopmental disorders, neurodegenerative processes, acquired brain injuries, psychiatric conditions). For example,

---

[25] See www.NavNeuro.com/17 and www.NavNeuro.com/29.

[26] Per the Taxonomy, a student must complete a minimum of two neuropsychology practica for neuropsychology to be considered a Major Area of Study.

we encourage adult-focused students to receive some neuropsychological experience working with pediatric populations and vice versa. We also recommend that every trainee try to complete at least one practicum that includes regular inpatient evaluations, as the clinical demands of gathering brain-behavior assessment data in this setting are invaluable to a neuropsychologist. Overall, this is the time in your career to dabble in clinical experiences that are outside of your specific interest area.

> Make sure every neuropsychology trainee has the opportunity to work with patients who are not 'testable' by our standard measures and to understand the types of things they can do to provide information about that patient and how that can be useful to the patient and their family.
>
> – Beth Slomine, PhD, ABPP-CN

> Be open-minded. I started my career thinking that I was going to focus on pediatrics and I ended up focusing on geriatrics. Had I not been open to experiences in something other than what I thought I was going to do, I wouldn't have found my passion, so keep an open mind.
>
> – Laura Lacritz, PhD ABPP-CN

> You want to get as many experiences as you can and get exposed to as many populations as possible… Later on in your training, you can specialize more.
>
> – William Barr, PhD, ABPP-CN

In order to achieve this lofty goal of working in so many different clinics and with so many different people, we recommend that you strive to have at least one practicum experience at an academic medical center. This setting tends to offer a great deal of clinical diversity, which could be one reason why internship and post-doctoral training directors have ranked it as highly desirable (Driskell et al. 2020; Ritchie et al. 2012).

As you move forward in your training, consider gradually narrowing your focus within neuropsychology.[27] Start thinking about which aspects of the field pique your interest – for example, do you enjoy working with Veterans with traumatic brain injuries, children with intellectual disabilities, or adults with multiple sclerosis? As you work toward targeting your area of subspecialty, you can begin aligning your research to be in this area, and you can pursue clinical opportunities that will further your experience and expertise in this niche. This process of moving from broad to narrow training can be visualized in Fig. 6.1.

> I don't think you should get too narrow too soon. One of the advantages of training is you can get a sense as to what opportunities there are in the field. If you have all your interest in one narrow world, it's a little dangerous. It's good to have a couple core ideas, and if one crumbles for some reason, you can press on with the other one. This is especially true early in one's research career.
>
> – Bruce Hermann, PhD, ABPP-CN

One important aspect of specialization within neuropsychology pertains to the age of our clientele because most neuropsychologists work exclusively with either

---

[27] Some people consider themselves general neuropsychologists and this is good for the field. We need a mix of specialty experts and generalists.

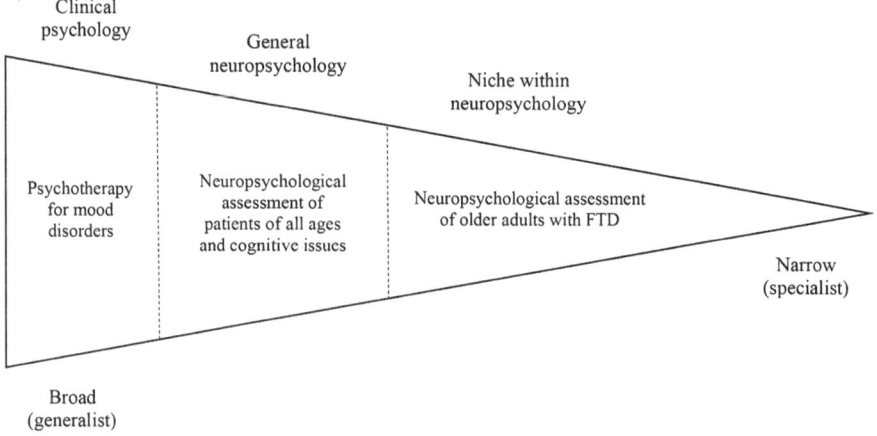

Fig. 6.1 Advancement of clinical training in neuropsychology, with examples of each stage of training

children/adolescents ("pediatrics") or adults and/or older adults ("geriatrics"). We (John and Ryan) both fall into the second camp, but many of our colleagues are pediatric neuropsychologists. We have learned from them that there are fewer opportunities to work with children compared to adults, so it is helpful for interested students to plan and search out graduate programs and practica that include at least some work with patients under the age of 18. If there are no such training opportunities available in your graduate program, one alternative strategy is to work with young adults with neurodevelopmental disorders such as ADHD, autism spectrum disorder, or learning disorders, and then apply to internships and postdocs with pediatric rotations.[28]

* * * *

As you learned in Ch. 1, there are many aspects to a comprehensive neuropsychological evaluation, including a review of medical records, a clinical interview, behavioral observations, test administration, scoring, interpretation, report writing, consultation with other providers, and a feedback session. Of course, trainees are not expected to have well-developed skills in these areas when they begin their first neuropsychological practicum – learning occurs along the way. Specifically, as the practicum year progresses, autonomy typically increases as the trainee gradually acquires additional skills. We believe that a reasonable goal is to be capable of independently managing most aspects of a neuropsychological evaluation (albeit with a supervisor's guidance) by the beginning of internship, and certainly by the beginning of a postdoctoral fellowship.

---

[28] If you want to specialize in pediatrics, we encourage you to look into the requirements for subspecialty board certification (https://theabcn.org/subspecialty-certification-in-pediatric-clinical-neuropsychology/; also see Chapter 7.

Table 6.3 Order of operations for neuropsychological training tasks

Part I
• Shadow others (supervisor, psychometrist, and/or upperclassman) as they conduct full neuropsychological evaluations. Follow-up with a debriefing and question/answer session. • Read the manuals for each test in your supervisor's core battery. • Practice standardized test administration with colleagues and/or supervisors. • Practice rescoring results from your supervisor's cases for fidelity, referring back to the manual as needed. • Create the score table.
Part II
• Draft the "Background Information" section of the neuropsychological report. • Administer tests to patients while being observed by supervisors and/or advanced trainees. • Independently score the results and ask supervisors and/or advanced trainees to check your work. • Review records and incorporate the information into the neuropsychological report.
Part III
• Administer and score tests independently, asking questions as they arise. • Conduct interviews and medical records reviews, first with your supervisor present and then independently.
Part IV
• Conduct full neuropsychological evaluations independently, asking questions as they arise. • Draft full neuropsychological reports.
Part V
• Teach parts I through III to early trainees.

There are several different models of clinical supervision, and each neuropsychologist will have their own preference as to how they approach the process of training a graduate student (Gauthier et al. 2020; Schwent Shultz et al. 2014; Stucky et al. 2010). We ascribe to the "see, then do, then teach" method, whereby students first observe the process, then take the reins, and eventually teach subsequent students. Table 6.3 shows our preference regarding an ideal order of training. This may conflict with your supervisor's method; in that case, we recommend that you follow your supervisor's guidance.[29]

There are many dimensions of neuropsychological training and we will not presume to do more than scratch the surface in this book. This is what graduate school and postdoctoral fellowship are for, after all. However, we will briefly discuss two topics that are essential parts of early neuropsychological training: test administration/scoring and report writing.

---

[29] We encourage you to read more about supervision of practicum students in the AACN practicum guidelines (Nelson et al. 2015) and use the form on competency benchmarks in the article's appendix with each of your practicum supervisors. The article also includes a list of competencies expected of neuropsychology trainees, specifying the degree of readiness expected for practicum and internship. We recommend that you familiarize yourself with those competencies (as well as those listed in Smith and CNS 2019, Heffelfinger et al. 2020, and Hessen et al. 2018) so you know what is expected of you.

Accurate test administration/scoring is the foundation of all interpretations, conclusions, and recommendations in a neuropsychological report. In other words, errors at this step compromise all other aspects of the evaluation. As such, it is profoundly important that you learn to administer and score tests correctly from the start, avoiding any bad habits that could arise from sloppy training (see https://www.NavNeuro.com/24). In addition to thoroughly reading test manuals, rehearsing test instructions, and role-playing with colleagues and/or supervisors (as both examinee and examiner), we recommend that you carry a printout of standardized instructions with you for each test, especially during this learning stage. This way you can read the directions verbatim and refer back to the details of each measure (e.g., start point, time limit, discontinuation rule). Additionally, we recommend noting every patient response meticulously and double-checking your scoring. In the beginning of your training, these demands will likely push the limits of your processing speed and working memory while in the room with a patient, but you will find that the process becomes much less challenging with practice. Once you have the standardized test administration down pat, you can begin focusing on building rapport with patients, observing and interacting with them in a fluid, attentive manner at all times.

Report writing is an even more complex task than is administration/scoring because it is more nuanced, intellectual, and abstract. Unlike administration and scoring, there is no standardized manual to teach you this skillset, and yet it is one of the most important functions served by neuropsychologists. We will not presume to be able to show you the ropes in a few sentences. Instead, we will encourage you to soak in your graduate school training and we will refer you to a well-regarded book, *Neuropsychological Report Writing* (Donders 2016), edited by Dr. Jacobus Donders, as well as two podcast episodes on the topic (https://www.NavNeuro.com/25 and https://www.NavNeuro.com/26).

Early Internship Preparation

Although internship applications are submitted in the later stages of graduate school training (see *Applying to Internship*, below), it is helpful to begin thinking about this task sooner rather than later. Importantly, neuropsychological internships are quite competitive and you will put yourself in a great position if you begin planning in advance. To be more concrete, one major aspect of "planning" is to track all of your clinical experiences, beginning with patient #1. We recommend this because you will be required to include a breakdown of all of your clinical experiences with your internship application. This means noting every face-to-face hour spent with a patient (even when shadowing someone else), not only noting the time but also the patient characteristics, the nature of the assessment or intervention, and any tests administered. In addition, you estimate time spent writing therapy notes and assessment reports, as well as hours spent in supervision.[30] This process can become unwieldy as your clinical hours increase, which is why we strongly recommend using software that organizes the data for you. There are several services available for this purpose such as *Time2Track*. This platform has multiple user-friendly features, such as the ability to schedule repeating events (e.g., "psychotherapy every Tuesday with patient X, a 38-year-old Asian-American woman"). Alternatively, some programs offer locally grown spreadsheets that translate directly to the internship application.

Given that internships will have access to this information, a common question is, "What will they be looking for in a competitive applicant?" Admissions committees examine every aspect of students' applications. This means that your undergraduate and graduate GPAs, research accomplishments (posters, publications, and presentations), letters of recommendation, essays, cover letters, etc., will all be scrutinized. But keep in mind that a psychological internship is a *clinical* experience, so your prior clinical work will be weighed heavily. In this vein, there are several aspects of your application that the admissions committees will prioritize. These include (1) the number of face-to-face patient hours[31], (2) a mix of assessment and intervention experience, (3) the sociocultural and clinical diversity of patients served, (4) the number of tests administered, and (5) the number of integrated reports completed. Within those categories, neuropsychologically focused internships tend to weigh clinical assessment experience most heavily (Ritchie et al. 2012).

Unfortunately, we cannot provide you with a gold standard criterion in terms of the number of face-to-face patient hours to accrue because programs vary widely in their requirements. However, we can share some data with you.[32] Ritchie et al. (2012) surveyed 75 supervising neuropsychologists from internship programs. In order to be considered competitive, supervisors expected applicants to have acquired a median of 400 practicum hours in neuropsychological assessment, with a mean of 575 hours (standard deviation = 506). Although not directly addressed in the article, we think that it is important for you to balance assessment and intervention (e.g., psychotherapy) because intervention hours are also critical to a well-rounded application.[33] Finally, internship sites prefer quality over quantity (Alden et al. 2000), so prioritize skill acquisition and choose practicum experiences that offer the best training.

****

---

[30] There are criteria that need to be met in order to count your hours. See detailed instructions through the following site: https://help.liaisonedu.com/AAPI_Applicant_Help_Center.

[31] This may be impacted by the COVID-19 pandemic.

[32] One caveat: these findings are nearly 10 years old at the time of writing, so the climate may be somewhat different by the time you are reading this.

[33] As we noted earlier, we have found that most neuropsychology-heavy internship programs prefer students who have generalist training in addition to assessment experience. We advise that you trumpet the fact that you have both skills in your essays and during interviews.

Before moving on, we want to expand on three clinical topics that deserve your attention: cultural diversity, test administration, and integrated reports. First is diversity. Throughout your practica, you will be recording many characteristics of the patients you serve. When the time comes to apply for internship, you will submit a data sheet with boxes that represent sociocultural and clinical intersectionality (e.g., the number of White homosexual male patients with depression, the number of Black heterosexual female patients with substance use disorder, and so on). No one will have worked with a large number of people of every background, but you will benefit from broadening your knowledge and skills by prioritizing variation across patients in every relevant dimension (e.g., age, sex, race, disability status, and religion). To do this, we recommend informing your supervisors about your interest in cultural diversity and then asking for the opportunity to work with people who are different from your prior clientele.[34]

Second, you will also need to track the psychological, neuropsychological, and psychoeducational tests used during graduate school (shadowing your supervisor counts). It is important that you at least gain familiarity with the most commonly utilized measures because they are the cornerstone of a neuropsychologist's armamentarium. In other words, it is best to have a broad knowledge of a wide range of different tests under your belt so that you can flexibly select the best test to administer to your patient, no matter their background and presenting problem. Rabin et al. (2016) list the most commonly used measures that we recommend you learn to administer, score, and interpret by the time you begin internship.[35] In many of your practica, you will use your supervisor's test battery by default, so you can broaden your horizons by working at a variety of different practicum sites. Additionally, within an individual site, many supervisors are open to suggestions in terms of additions to their battery.

Finally, we have assessment reports. These must be "integrated," meaning that you have consolidated data from multiple sources.[36] For applicants to neuropsychology-focused internships, these reports are valued greatly because they represent an advanced, high-level product that is similar to the work of clinical neuropsychologists on a daily basis. Similar to our advice about publishing peer-reviewed manuscripts, the number of integrated reports written by neuropsychological

---

[34] For more information on cultural neuropsychology, check out the following resources: AACN Relevance 2050 (https://theaacn.org/relevance-2050-initiative/), the Society for Black Neuropsychology (https://www.societyforblackneuropsychology.org), the Hispanic Neuropsychological Society (https://hnps.org/), the Asian Neuropsychological Association (https://www.the-ana.org/), ANST webinars (https://scn40.org/ema-webinars/), and several NavNeuro episodes on cultural neuropsychology (e.g., www.NavNeuro.com/21 and www.NavNeuro.com/58).

[35] Students interested in working primarily with children do not need an in-depth understanding of adult measures, and vice versa.

[36] Per AAPI Online's guidelines: "The definition of an integrated psychological testing report is a report that includes a review of history, results of an interview and at least two psychological tests" (https://www.appic.org/Internships/AAPI/Integrated-Report).

internship applicants varies widely (in our experience, from 20 to 100) and the best advice is, "the more high-quality reports, the better."

Applying to Practicum Sites

Applying to practicum sites will be similar to other school and work applications with which you are now familiar. You will first identify a list of available options by asking the administrative staff, your advanced colleagues, and the clinical faculty. Ideally, several neuropsychological sites will be available and you can rotate through each of them. If not, you can blaze your own path with some creativity and hard work. For example, consider reaching out to private practice neuropsychologists in your area to inquire about whether they are interested in taking on a practicum student.[37] If you take this route, your doctoral program will need to create a formal written affiliation agreement with the training site. To do this, talk with your DCT and refer to the American Academy of Clinical Neuropsychology (AACN) practicum guidelines (Nelson et al. 2015) for relevant policies and procedures.[38]

> Students that hustle get a lot of respect in my book. I've had internship applicants who've said, 'my program didn't have neuropsychological training but it was something I wanted so I went out and created a practicum.' Awesome! You get bonus points for that! That really shows me that you're trying to get that relevant experience.
>
> – Suzanne Penna, PhD, ABPP-CN

****

Earlier, we mentioned the importance of obtaining diverse neuropsychological experiences. Now we will provide several additional tips in the context of decisions about practicum sites.

First, it is advantageous to work under a board-certified neuropsychologist; you can identify such a person by looking for "ABPP-CN" or "ABN" after the "PhD" or "PsyD" in their signature line. Board certification confers prestige to both the recipient and their collaborators/trainees. Many high-quality neuropsychologists are not board-certified, but this is the highest clinical credential in the field, so it is a great shorthand for advanced competencies (see Ch. 7). Second, seek out a training site that is nationally known. Not everyone has UCLA or Harvard in their backyard, so it is understandable if you do not have a "big name" on your curriculum vitae (CV). However, if you can go the extra mile (literally and figuratively) to train at one of those sites, then do it. It will benefit you to be associated with a highly reputable

---

[37] To find them, ask your supervisors and/or Google "neuropsychologists in my area."

[38] If you would like to supplement the neuropsychological mentorship you receive from your graduate program/practica, consider participating in the AACN Student Mentorship Program (https://theaacn.org/student-mentorship-program/).

Table 6.4 Student/trainee neuropsychological groups

Parent organization	Student group	Website
International Neuropsychological Society (INS)	Student Liaison Committee (SLC)	https://www.the-ins.org/about-ins/ ins-committees/ student-liaison-committee-slc/
Society for Clinical Neuropsychology (SCN)	The Association of Neuropsychology Students and Trainees (ANST)	https://scn40.org/anst/
National Academy of Neuropsychology (NAN)	NAN Student & Post-Doctoral Resident Committee (NANSPRC)	https://www.nanonline.org (hover over the "Professional Resources" tab and click "Student & Trainee Resources")
American Academy of Clinical Neuropsychology (AACN)	Student Affairs Committee (SAC)	https://theaacn.org/students/

institution. Of course, quality should always trump name recognition, so if there is a discrepancy between the two, choose the former.

Most sites will ask for a CV and will conduct an interview. Some will ask for several letters of recommendation, a list of tests you have administered, the number of intervention/assessment hours acquired, and other materials. To help with this process, see our advice for graduate school applications in Ch. 5 and internship applications, below.

Note: If you have already completed several years of graduate school and have now decided to pursue neuropsychology, we have a few pieces of advice. First, consider staying in graduate school for one extra year in order to beef up your neuropsychological training as much as possible. A year may seem like a long time, but if it is financially feasible, building up more training for your CV will help a great deal on internship applications. Second, immediately work toward acquiring as much neuropsychological practicum experience as possible in terms of clinical hours, integrated reports, and multiple practicum sites. Third, become involved in student organizations in neuropsychology (see Table 6.4). Finally, ask for advice from neuropsychologists in your graduate program. They are the best people to help you find more neuropsychological experience in your area, and they may also have an idea as to where you could apply for internship.

Action steps:

 Acquire general psychology experience in early practica.

 Acquire general neuropsychology experience in later practica.

 Track your intervention and assessment hours and patient characteristics beginning with patient #1.

 Learn and fine-tune your basic neuropsychological skills such as test administration/ scoring and knowledge of the psychometric properties of tests.

Networking and Professional Development

I was really shy as a trainee and afraid to speak up and get to know people. I wish someone would have told me to spend more time working on my networking skills.

                                                                                    – April Thames, PhD

The term "networking" sometimes carries with it a negative connotation because people use it to refer to a means of unilaterally securing favors from those in power. It is related to ideas such as "workplace politics" and "nepotism." This is not what we mean when we use the term. Rather, we are referring to the process of building mutually beneficial and authentic relationships with other people in neuropsychology. The benefits of these connections are numerous. Not only does networking lead to a more tightly knit, closely connected field on a broad level but it also keeps individual professionals abreast of new findings, techniques, and innovations, and it enhances collaboration amongst clinicians and researchers alike.

Students have a great deal to gain from becoming well-connected, and graduate school is the ideal time to focus on this area of professional development. Given the diversity of settings and specialties within the field, the more neuropsychologists and trainees you know, the more knowledge and avenues will be available to you. Additionally, you will be in need of many favors over the next several years, from the composition of letters of recommendation, to the editing of application materials, to advice on how to handle difficult ethical situations, to consultations on challenging clinical cases. And even more important than favors is the satisfaction and fulfillment that comes with spending quality time with friends and colleagues. We have both found that our most gratifying professional experiences involve positive interactions with our fellow neuropsychologists.

Of course, building relationships is about giving as well as receiving, and offering value to others is about putting yourself out there and showing that you care about their careers as well as your own. This can be accomplished in many different ways, both at your institution and in the field broadly. At your university, we suggest contacting neuropsychologists who share your interests. Ask to schedule a meeting with them, attend their office hours, sit in on their lectures, and/or attend thesis/dissertation defenses where they are committee members. Even if they are not your research advisor, ask about their ongoing projects and express interest in their work. You might end up collaborating with them on manuscripts. Most faculty members will greatly appreciate this approach and it will allow you to enrich your professional network.

There are also many opportunities for networking and professional development outside of your university. The lowest hanging fruit is attending neuropsychological conferences where you can meet like-minded people, showcase your research, and attend lectures and events. We listed the primary neuropsychological conferences and their websites in Ch. 5 (Table 5.2). We strongly encourage you to join these organizations and attend as many of their meetings as you can. Make it your goal to network as much as possible during the conference. There are several ways to do this:

1. Simply walk up and introduce yourself to people whose work interests you.
2. Mingle at poster sessions, asking a variety of people about their research.
3. Attend social events and student activities.
4. Attend lectures and talk to other attendees and/or the presenters afterward.
5. Email conference staff a few months in advance and volunteer to help – this may involve working the registration desk, setting up rooms, and/or packing tote bags for attendees.[39]

In addition to conferences, you can also join student groups and committees. All neuropsychological societies have some form of student outreach in which you can participate. Table 6.4 includes a list of some of the most well-respected organizations available. Many faculty-level committees and boards also have a seat saved for a student member, so keep a lookout for available positions related to an area of interest. For example, the INS Student Liaison Committee (SLC) has openings for student representatives. Additional options include joining your state or provincial psychological or neuropsychological association, signing up for neuropsychology-related listservs (https://scn40.org/anst/other-listservs/), and/or following one of the neuropsychological organizations' social media feeds.

Social media can be a helpful tool for networking and professional development, not to mention staying connected with friends and family. Consequently, we recommend that you consider how you might use platforms such as Facebook and Twitter to engage with the neuropsychological community. At the same time, we would be remiss if we did not mention a potential downside or pitfall of social media use. That is, it is easy to lose large chunks of time checking and rechecking social media feeds, as they can monopolize the *attention economy* of our lives,[40] and the last thing you want in graduate school is to waste time. Consequently, many people (us included) choose to limit social media use in order to maintain mindful attention and increase productivity.

Action steps:

 Introduce yourself to professors at your university whose work is interesting and relevant to neuropsychology.

 Join at least one neuropsychology-related organization and attend the annual neuropsychological conferences.

---

[39] For more networking strategies, listen to the following podcast episode: https://www.choosefi.com/networking-with-jordan-harbinger-ep-233/.

[40] See the work of Tristan Harris: e.g., https://www.wired.com/story/our-minds-have-been-hijacked-by-our-phones-tristan-harris-wants-to-rescue-them/; https://samharris.org/podcasts/218-welcome-cult-factory/.

Apply for a position in a committee (e.g., in a neuropsychological and/or state organization).

 Join at least one neuropsychology-related listserv.

 Be mindful of your attention and minimize distractions.

Teaching

Similar to networking and professional development, teaching is an important area of focus in graduate school. Students vary in their interest in teaching and in their plans to incorporate it into their careers, but even for those of you who do not plan to teach consistently for the next few decades, some experience in this domain is still beneficial to your growth and development because pedagogical skills will generalize to other areas of your professional life (e.g., clinical supervision, colleague consultation, conference presentations, patient psychoeducation). Consequently, we recommend that you at least dabble in teaching during graduate school and, if your goal is to seek an academic, student-oriented career, that you begin building the teaching portion of your CV. Fortunately, you will likely have a plethora of opportunities to do this during the next 5–6 years. You can work as a teachers' assistant (TA) for neuropsychology-related courses, you can offer to guest lecture in professors' classes and/or at departmental seminars (sometimes called "brown bag lunches"), and you can seek out undergraduate courses to teach independently.[41] If you elect to serve as an instructor of your own class, keep in mind that it requires a great deal of time and energy, so it is important to be thoughtful about your schedule and how you can fit it into your own coursework, as well as clinical and research responsibilities. Regardless of how much teaching you engage in, and in what context, we recommend that you both note each of these experiences on your CV and hand out teaching evaluations at the end of class periods or lectures so that you have data when you apply for a job later on. We have included a sample at www. NavNeuro.com/book.

---

[41] A helpful resource for prospective teachers is *McKeachie's Teaching Tips,* 14th edition (Svinicki and McKaechie 2014) by Marilla Svinicki and Wilbert McKaechie.

Comprehensive Examination

All clinical psychology programs require students to pass some form of a comprehensive examination (aka "comps," "prelims," or "the qualifying exam") as part of the degree requirements.[42] The format and difficulty of this test varies greatly by program, so it is impossible for us to provide you with specific guidance. One general tip is to speak with advanced students about their experiences and ask for their advice on how best to prepare. Often, study strategies include reviewing notes and textbooks from your core courses, rereading key articles assigned by professors, and participating in study groups. Importantly, some clinical psychology programs require students to take the Examination for Professional Practice in Psychology (EPPP) – the licensure examination – to fulfill the requirement for a comprehensive examination. See Ch. 7 for more information about the EPPP.

Managing Finances

The financial toll of graduate school varies widely depending on the specific program. You may be provided with full tuition remission and you may even receive a stipend on top of that. On the other hand, you may be borrowing thousands of dollars in loans each year in order to fund your tuition and personal expenses. Either way, we encourage you to live frugally and minimize the amount of money borrowed. This may seem obvious, but we have found that many students overlook the magnitude of both the loan principal and interest. Ignoring these facts is particularly precarious if you have unsubsidized loans (meaning that interest begins accruing as soon as you borrow it) or loans with a relatively high interest rate (e.g., Graduate PLUS loans, which carried a 7.08% interest rate for the 2019–2020 school year). So, the first bit of advice is to know (1) the type of loan accepted and (2) the charged interest rate.

You will be happy to know that all APA-accredited internships pay a stipend, and postdoctoral fellowships do so as well. In fact, a general rule of thumb is that interns earn a low wage, around $20,000–$35,000 for the year, then the salary nearly doubles during fellowship years (roughly $50,000), and then it doubles again for a first professional job (roughly $100,000).[43] Given the lower wage during internship, some people do borrow money during this year. However, unless you are in a high cost of living area (e.g., Palo Alto or New York City), or have extenuating circum-

---

[42] Some programs allow for non-study options such as writing a predoctoral grant or a systematic review.

[43] See the latest "Salary Survey" (Sweet et al. 2020a, 2020b) for more details about average postdoc and early career salaries.

stances (e.g., children), it is possible to get through internship without acquiring more debt.[44]

We covered the average cost of graduate school, provided an overview of different types of loans, and recommended several strategies for minimizing debt in Ch. 5, so we suggest that you go back and read that section if you have not done so already.[45] In addition to that information, we will share a few ideas about obtaining funds that are unique to graduate students.

Generally speaking, many universities offer financial compensation to TAs and research technicians. We recommend that you ask your program's administrator early in graduate school about the available opportunities for funding, especially if you can benefit academically from the work itself (e.g., research and teaching). Many universities also have part-time job opportunities that offer down time for working on academic tasks. For example, I (John) served as the assessment librarian at my university. Students would occasionally check out a testing kit from me, and the rest of the time I earned money while simultaneously writing my dissertation. If you are not able to find paid work within the purview of your specific graduate program, you can also ask the Student Resources department and the grant/fellowship office at your university whether they are aware of any graduate school funding sources.

Outside of the university, there are many grants, awards, scholarships, and other sources of funding that we encourage you to look into and apply for. The APA lists over 600 sources of graduate school funding on their website (https://www.apa.org/about/awards/index).[46] Many of these opportunities are specific to women, people of color, and other underrepresented groups.[47]

In terms of neuropsychology, most of the major organizations offer research grants and awards. For example, SCN has an annual dissertation award (https://scn40.org/eac/) and INS offers annual research and travel awards (https://www.the-ins.org/about-ins/ins-awards/).[48] We strongly encourage you to apply early and

---

[44] Keep in mind that you will need to budget for the internship acquisition process, both in fees and travel costs. In 2018, applicants reported mean total costs of $2,323 (SD = $1,804; Keilin 2018).

[45] We also cover information regarding finances/student loans in Chapter 7.

[46] Also, see the following sites for links to other funding sources, such as the National Science Foundation Graduate Research Fellowship Program: https://www.apa.org/gradpsych/2016/01/research-funding-sources; https://www.apa.org/education/grad/funding; https://www.psichi.org/page/awards#graduate.

[47] As noted in the Preface, there are multiple psychology-related resources specifically designed to assist people of diverse backgrounds. A few include: https://www.apa.org/apags/resources/; https://www.apa.org/pi/disability; https://scn40.org/piac-ema/; https://www.nanonline.org (click About NAN > NAN Committees > Culture & Diversity Committee); Queer Neuropsychological Society (https://www.queerneuro.org.). We include more links at www.NavNeuro.com/book.

[48] The Association of Neuropsychology Students & Trainees (ANST) includes a list of other awards and resources: https://scn40.org/anst/funding/

often for these awards. Not only does winning an award typically come with a cash prize, but it is an excellent addition to your CV.

Finally, to add to the resources available in your department, university, and the field of neuropsychology at large, there are also books, podcasts, and blogs on financial management that are not specific to graduate school, but that can be useful to graduate students. Of course, everyone's financial situation is unique, and we will not presume to serve as your personal financial advisor. But we do want to share knowledge with you, in the event that you are interested in learning more about personal finance. If so, see Ch. 7.

Applying to Internship

Overview

The final year of graduate school consists of a formal internship,[49] which is one year of full-time work at a new institution, separate from your graduate program.[50] The internship is primarily a clinical experience, with the majority of the time spent on relevant patient-focused tasks and the remainder spent on professional development, didactics, and potentially a small amount of research. Programs vary in their requirements for allowing students to begin internship. Typically, you must have (a) completed all coursework, (b) acquired sufficient clinical experience through practica, (c) passed your comprehensive examination, and (d) proposed your dissertation. This just leaves the dissertation defense, and programs allow this to take place during the internship year. Check with your institution well ahead of time to clarify their expectations and deadlines.

Now, even though it is possible to leave for internship prior to finishing the dissertation, it is very helpful to have defended one's dissertation beforehand. This is one of the reasons why we emphasized in the *Thesis and dissertation* section that you select a dissertation topic that is well-contained and manageable. Internship is busy and demanding in and of itself, and it only adds time and complexity to the mix if you are forced to juggle current responsibilities with a massive research project with collaborators in another city. For example, if the project is ongoing, you may still be actively collecting data, which is very difficult to coordinate from afar. Even if data collection is complete, you will be required to schedule ongoing meetings with your advisor and committee members about the status of the project, and you may be required to fly back to your graduate institution to defend the dissertation in person. Finally, most people have not defended their dissertation prior to the start of internship, so if you are able to do so (or to at least schedule a defense date), you

---

[49] Internship is sometimes referred to as "residency," although this can be confusing because some refer to postdoctoral fellowship as "residency."

[50] Some graduate programs also have internship programs.

will have a competitive advantage during internship applications and interviews. In other words, internship directors and faculty supervisors look favorably upon students who get ahead of the curve by defending their dissertation early (Ritchie et al. 2012).

* * * *

Completing internship applications is a time-intensive, challenging task that requires a high degree of effort and organization. We recommend that you think about this as a long-term, multistep *process,* and begin planning for it early (see below). The psychology internship is meant to significantly increase your clinical, professional, and research skills as a budding clinical psychologist, and it will set you up for a good fellowship experience and for future employment.

The materials required for the application will likely be familiar to you; you will provide academic transcripts, a data summary of your clinical training, your CV, three letters of recommendation, a cover letter tailored to each site, and four 500-word essays. Because you will be applying to sites with neuropsychology tracks, some programs will also require one or two sample integrated reports. We will briefly cover all of these materials below, but we strongly recommend that you read the book *Internships in Psychology: The APAGS Workbook for Writing Successful Applications and Finding the Right Fit* (Williams-Nickelson et al. 2019, 4th edition).[51] This book provides an in-depth discussion of each aspect of the application process, and provides sample essays, CVs, and other helpful materials. And although not a substitute for the Workbook, we also recommend a podcast episode on this topic (www.NavNeuro.com/53).

Timeline

Before diving into the materials and site selection, we recommend that you visit the Association of Psychology Postdoctoral and Internship Centers (APPIC) Application for Psychology Internships (AAPI) site and create an account (https://www.appic.org/Internships/AAPI). You will use the AAPI Online service to upload all of your application materials, so it will be to your benefit to familiarize yourself with it sooner rather than later. Most applications are due in late October through mid-November, so we recommend beginning to seriously think about the application process early in the summer of the year you plan to apply (see Table 6.5).

---

[51] Check out these webinars by the authors: https://www.youtube.com/watch?v=hyg_I4u8tC8, https://www.youtube.com/watch?v=3ZR7cGxLD3o&feature=youtu.be.

Table 6.5 Sample internship application timeline with preparation steps

Month	Steps
May–June	Subscribe to Match-News, the APPIC email listserv.[a]
	Draft essays.
June–July	Familiarize yourself with the AAPI website and create an account.
	Create a spreadsheet to organize information about sites and begin filling it in.
	Draft a cover letter for a site to which you know you will apply.
	Request letters of recommendation.
August	Update your CV and ask mentors/faculty to review it.
	Continue drafting essays and the cover letter; ask faculty to review them.
	Follow up with supervisors regarding letters of recommendation
	Request that official graduate transcripts be sent to AAPI.
September	Finalize essays.
	Draft and edit cover letters tailored to each site
	Finalize your tracked clinical experiences.
	Locate and scrutinize sample reports (if necessary).
October	Register for the match.[b]
	Finalize cover letters.
	Submit applications.
November	Prepare for interviews by writing down responses to common questions and by role-playing with friends/mentors.
December–January	Complete interviews.
February	Submit and certify your rank order list prior to match day.

[a]https://www.appic.org/E-Mail-Lists/Choose-a-news-list/Match-News
[b]This can be done up until 12/1, but it is helpful to include your registration number with your applications. Register at https://natmatch.com/psychint/

Required Materials

Essays As part of the application, you are required to submit four essays, each of which can be no longer than 500 words. The prompts are the same for all applicants: there is an autobiographical statement, a description of the trainee's theoretical orientation, a discussion of experiences working with diverse populations, and an explanation of research interests and accomplishments. We suggest that you refer to the APAGS Workbook for in-depth advice, including several samples of each essay. Although you can probably write 500-word essays in your sleep by now, these need to be powerful and polished, and the process requires significant time and energy. This is also a different style of writing than is required in clinical and research contexts – it is more creative, personal, and engaging. Thus, we strongly encourage you to begin these essays months prior to the deadlines and to have them vetted by multiple colleagues and supervisors; be prepared to work through multiple drafts of each document. In addition to weaving your career goals throughout the essays, it is important to balance novelty and creativity (e.g., anecdotes from your childhood) with conventionality (e.g., "CBT!" or "research informs clinical practice!"). In other words, you want to stand out in a good way.

Letters of Recommendation We also provided general advice in Ch. 5 regarding requesting letters of recommendation, so we suggest that you reread that section. Specific to internship applications, all programs require three letters and some sites allow four. Each of your letter writers will complete a Standardized Reference Form directly on the AAPI. You will be notified when each supervisor has completed the letter, but you will not be able to open the document itself. The Standardized Reference Form includes prompts such as the capacity in which the supervisor worked with the trainee, the duration of the professional relationship, the activities completed under the supervisor's guidance, and the trainee's interests, aspirations, professional development, competencies, and areas of growth.[52]

In terms of who to approach about writing a letter, we recommend that you ask your primary research advisor and two other supervisors who know you well. You want your letter writers as a whole to speak to every aspect of your graduate school performance (research, clinical work, teaching, etc.). Moreover, it is typically preferable for these individuals to be board-certified neuropsychologists, but a non-neuropsychologist (e.g., a clinical psychologist or cognitive neuroscientist) can be a great selection if you have strong a relationship with them. Consider the sites to which you will be applying and choose letter writers accordingly. For example, if several of your top sites include health psychology rotations and if you have worked extensively with a health psychologist, then ask that person to write you a letter.

CV See Ch. 5 for general advice regarding your CV. The overall format will likely remain the same, but you will add new experiences from graduate school such as research activity and output (e.g., publications, presentations, grants/awards), clinical training, teaching, professional service, and references.

Cover Letters You will include a 1–2 page cover letter tailored to each site to which you apply. This is often the first application item to be reviewed, so we recommend that you allocate sufficient time to crafting well-written letters that pique the reviewers' interest as they prepare to look through the rest of your materials. Generally speaking, the cover letter can be divided into two parts – part one (1–2 paragraphs) focuses on your training and experiences, and part two (3–5 paragraphs) discusses aspects of the site that are of interest to you and that fit well with your background. Part one need not change much from site to site (minor alterations will highlight portions of your training that fit best with the program in question), while part two will differ drastically across sites. When writing part two, we suggest that you reference specific clinics by name and specific clinicians/researchers with

---

[52] We recommend that you frequently ask supervisors for feedback about areas of growth throughout your training so that you can address any potential issues well in advance. For example, you may open the door by saying, "I am always open to constructive feedback about my work – is there anything that I can improve upon?"

whom you would like to work. Make it clear that your past experiences, gaps in training, and internship/career goals all align with the site's mission and offerings. Importantly, let your enthusiasm show. Similar to the essays, you want to stand out in a good way.[53]

Summary of Doctoral Training As mentioned above, you will submit a detailed accounting of your clinical experiences. *Time2Track* provides printouts of your hours that mirror the AAPI format to make life easier. If you have been diligently tracking your clinical experiences, then this will be a straightforward and easy process. If you have not, then it can be a lengthy and tedious undertaking.[54]

Sample Reports Sites can request up to two work samples, which include case summaries or testing reports. Given that you will be applying to sites with a strong neuropsychology focus, many will ask for a sample assessment report or two. It is important to discuss case selection with your advisor. Ideally, the report you choose will reflect a case that demonstrates your ability to administer, score, and interpret a typical neuropsychological battery of tests, as well as your proficiency in case conceptualization. This should be a case for which you completed a considerable portion (or all) of the evaluation and writing. Typically, the best sample reports are at a moderate level of complexity. In other words, the patient showed some degree of cognitive impairment (as opposed to a "worried well" person), but the case was not overly complicated or controversial. If the site requests two reports, try to select cases that are significantly diverse (e.g., a different etiology and/or age group). Importantly, be sure to remove all identifying information (e.g., name, birthdate, cities) from the documents.

Selecting Sites

The process of identifying sites and deciding where to apply will resemble the graduate school application process, and you will benefit greatly from a highly organized system. To this end, we recommend that you keep an evolving excel sheet that lists the sites you are considering, as well as the application due dates, rotations offered, number of positions available, strengths/limitations, etc. We provide a free excel sheet on our website to help get you started (www.NavNeuro.com/book).

In terms of requirements, the HC guidelines specify that "internships must be completed in an APA or CPA approved professional psychology training program" (p. 4). Although the guidelines are technically aspirational, we strongly recommend that you apply only to programs that are APA or Canadian Psychological Association

---

[53] The APAGS Workbook includes a sample cover letter.

[54] If you have not been tracking your hours, you can consult the AAPI instructions for how to proceed (https://www.appic.org/Internships/AAPI). We also recommend that you discuss the issue with your advisor and DCT.

Table 6.6 Degree of importance of internship site factors

Very important
• APA or CPA accredited
Important but not required
• At least one supervisor with board certification in clinical neuropsychology • Multiple, diverse rotations offered • At least 50% of time spent in neuropsychological activities
Applicant-specific priorities
• Geographic location • Type of institution (VA hospital, psychiatric facility, etc.) • Availability of a particular clinical population • Protected time for research • 9–5 schedule[a]

[a]On average, VA hospitals are more likely to offer a structured 9-5 schedule, while academic medical centers are more likely to have varied work schedules

(CPA) accredited.[55] Aside from accreditation, other internship variables rest on a spectrum of importance (see Table 6.6), and the HC guidelines offer a great deal of flexibility; we address a few relevant issues below.

****

One question that comes up frequently is, "What percentage of my internship time should be devoted to neuropsychological training?" There is no hard and fast rule, but several prominent groups have suggested that internship consist of at least 50% clinical neuropsychology training (e.g., INS-Division 40 Task Force 1987; Smith and CNS 2019).[56] Although ABCN currently does not require neuropsychology training during internship, postdoctoral training directors list "intensity of neuropsychological internship experience" as the most important factor for a competitive applicant (Driskell et al. 2020).[57]

Now, if you have undergone little generalist training, we think that it is in your best interest to select a site with several non-neuropsychology rotations. We have already explained *ad nauseum* why we believe that this generalist experience is essential for a neuropsychologist, and internship might be your last chance to acquire it. Importantly, this advice is very much in line with the HC guidelines, which state, "The purpose of the internship is to complete training in the general practice of professional psychology and extend specialty preparation in science and professional practice in clinical neuropsychology. The percentage of time in clinical

---

[55] As noted elsewhere, many fellowship programs, licensing boards, and employers only accept applicants who have completed accredited internships.

[56] Per the Taxonomy, in order for neuropsychology to be considered a "Major Area of Study" at an internship site, it must consist of at least 50% training in clinical neuropsychology and have didactic experiences consistent with the HC guidelines. Note that supervised clinical activities completed as part of research count toward this percentage.

[57] Mean ideal proportion of time spent by a competitive applicant in clinical neuropsychological activities on internship was 61.5% (SD = 13.6). Mean minimum proportion was 40% (SD = 13.8).

neuropsychology should be determined by the training needs of the individual intern" (Hannay et al. 1998; p. 4). To be specific, we recommend that, for most neuropsychology-focused applicants with three or four available rotations, two to three be neuropsychology-focused and one to two be generalist. Of course, your individual training needs should be the primary guide regarding the optimal degree of neuropsychological experience on internship.

Other common questions asked by students pertain to specific details about sites, such as the type of setting (e.g., academic medical center vs. rehabilitation clinic), the patient population, the number of supervisors, and the number of interns accepted into the program. These details typically do not make a major difference in the overall quality of training, and the ideal selection within each category is specific to individual applicants. However, there are a few details that can make a difference in the long-term:

1. It is preferable (but not required) for at least one of the supervisors at the site to be board certified in clinical neuropsychology.
2. If you are interested in a career in a Veterans Affairs (VA) hospital, then it is helpful to complete at least a portion of your training at a VA. This training can occur at any point in your trajectory (i.e., graduate practicum, internship, fellowship).
3. Most people will benefit from working in a broad range of settings with diverse clinical populations. Thus, you may consider prioritizing consortiums, which offer rotations in a variety of settings.
4. Although it does not matter where in the US you complete internship, casting a wide net will likely increase your chance of matching to one of your top sites. We understand that some people are geographically restricted for personal and/ or financial reasons – we simply want you to be aware that this can decrease an applicant's chances of matching to an internship.

****

Now that you are armed with general information about internship sites, it is time to contemplate narrowing down the field. Given that there were 642 APA-accredited internship programs as of 2020 (https://accreditation.apa.org), this process may initially feel overwhelming. However, you can substantially reduce the initial pool of potential options by ruling out sites that do not offer multiple neuropsychology rotations. The most straightforward method for doing so is to search directly on the AAPI Directory, refining your search by "neuropsychology" (https://membership.appic.org/directory/search). This will provide the most up-to-date information and will show many relevant details such as accreditation status, stipend, and application deadline. Another directory that we encourage you to take advantage of is hosted by SCN (https://scn40.org/training-directory/). Keep in mind, however, that this is not an exhaustive list and may not be updated, so cross-reference it with the APPIC Directory.

In addition to directories, we recommend asking for internship site selection advice from former graduate students in your program who completed the application process in the last few years. These people can be a great resource because they recently spent a large amount of time and energy contemplating the pros and cons of dozens of programs. Given that you both attended the same graduate program, their neuropsychological training will overlap to a large degree with yours and their advice will likely be applicable to your situation.

Even if you fully utilize search engines and friends/colleagues, you will still find yourself spending a lot of time reading about individual programs on their websites. Internship programs have a few ways of signaling that they offer neuropsychology training. For example, brochures or websites will often specify that they "abide by the HC guidelines," "adhere to the criteria developed by Division 40" (referring to the HC guidelines), or something similar. Also, many programs offer a neuropsychology track or concentration, meaning that much of the interns' time is spent in neuropsychology rotations, while the remainder is spent in other clinical areas (e.g., severe mental illness, health psychology, substance abuse). Attending one of these programs offers some assurance that interns will receive strong neuropsychological training. That being said, many excellent neuropsychology programs do not have a specific concentration/track, so you should thoroughly vet the directories that we provided earlier.

On the topic of narrowing down one's list of internship sites, many students wonder, "How many programs should I apply to?" Well, data from the 2018 APPIC applicant survey showed that (a) the mean number of applications submitted per student was 15.4 (standard deviation = 4.8; Keilin 2018) and (b) the match rate levels off at around 15. In other words, applying to more than 15 sites does not improve a student's chance of matching. Moreover, the application cost increases after 15 and, of course, every additional interview requires time and money as well. Consequently, some people recommend applying to no more than 15 total sites. In our view, 15 is a good benchmark, but the situation is nuanced and there may be a rationale for some people to consider applying to upwards of 20 sites. First, while 15 applications may provide you with an excellent chance of matching somewhere, you may not match with a program that is an ideal fit for you, and the fit is critical to maximizing the internship experience. Second, the interview process is a great way to increase your neuropsychology network, so additional interviews are not wasted time and money. Taken together, we typically recommend that most neuropsychology-focused graduate students consider applying to 12–20 internship sites.

You will be happy to know that match rates have improved dramatically over the past decade. Compared to the 2012 match rate of 74%, 2017–2020 match rates were ~90% (95–96% if you include Phase 1 and 2; APPIC 2020). Still, there are several factors to consider when selecting sites and judging your own competitiveness. First, look at the number of applications the site received in recent years compared to the number of available positions. Second, ask colleagues and supervisors about each site's reputation with respect to the factors of highest import to you (e.g., protected research time, work-life balance, and clinical diversity). Finally, consider the relationship between your graduate program and the internship program (i.e., how many students have matched at that site in the past), as internship admissions committees often feel more comfortable accepting students from graduate programs with which they are familiar. Once you have this information, we suggest diversifying your portfolio by applying to several sites that could be labeled as "a reach" (your chances of matching are low), several sites that could be labeled as "a sure thing" (your chances of matching are high), and a larger number of sites for which you are moderately competitive.

Note: Many students also selectively choose internship sites that have an open fellowship position for the following year. See the *Selecting Sites* section below for a discussion of the pros and cons of completing internship and fellowship at the same site.

Interviewing

In terms of general interviewing advice, we encourage you to listen to episode 8 of NavNeuro (www.NavNeuro.com/08). We link to many resources in the show notes for that episode, including tips related to travel, sample questions (both to ask and to prepare for), and much more. Also, see Ch. 5 for additional interviewing suggestions and see the *Self-care and stress management* section in this chapter for suggestions on how to manage anxiety and stress. In the meantime, we will provide information that is specific to internship interviews.

Traditionally, internship interviews have been conducted in person, but some sites have allowed for telephone or videoconferencing formats as well, and we predict that this will increase following COVID-19. You will likely begin receiving these invitations in mid-to-late November. A few sites will hold interviews in December, prior to the winter holiday, while most will occur in January. Of course, you will aim to maximize interview invitations, but keep in mind that crafting a logistically feasible itinerary can be challenging if you receive a large number of offers (in 2018, the mean number of offers received was 7.4 [standard deviation = 3.8]; Keilin 2018). If this is the case, there are two strategies to consider. First, you can turn down interview offers – we recommend doing so only for sites that are lower on your rank-ordered list and only after you have a full schedule. Second, you can reschedule with your less preferred sites if a highly ranked site offers you interview dates that are already taken. Regardless of the number of invitations you receive, we recommend scheduling your top sites later in the interview season (mid-to-late January) if possible, so that earlier interviews can serve as exposure and practice.[58]

---

[58] We also recommend role-playing with family, friends, and/or mentors multiple times prior to your first interview.

Once your schedule is finalized and you are preparing to meet, greet, and chat with supervisors and students at each program, it is time to dig deeper into the interviewing process itself. In that vein, we want to highlight a few important points. First, it is crucial that you have prepared a brief, well-oiled response to the statement, "Tell me about your dissertation." In our experience, this is the most common question asked by potential supervisors. As such, we recommend that you prepare a canned answer that is around 30–60 seconds long and includes the following components:

1. The genesis of your research question (including the relevant gap in the scientific literature).
2. The importance of this question.
3. A very broad overview of the study methodology.
4. One or two key hypotheses and associated findings.
5. A powerful ending statement reiterating the importance of the project.

So, you will think through your project in these terms and practice the summary of your dissertation. When you are in the room with an interviewer, you will deliver the summary and then ask the person if they have any follow up questions. Be prepared to go in any direction that they may take you. Also, keep in mind that it is acceptable if the project is not yet complete. In that case, explain the current state of the study and your planned defense date. Importantly, neuropsychology interviewees sometimes worry that their dissertation will be devalued if it does not sound like a traditional neuropsychological study. If this applies to you, then we recommend that you contemplate how and in what way your project is relevant to neuropsychology. To be clear, we are not suggesting that you lie or misrepresent yourself in any way, but rather that you extrapolate from your study to your broader field of interest. As an example, if you are working with animal models of fetal alcohol spectrum disorder (FASD), you are gaining an understanding of the underlying biology of that condition and its behavioral effects. You could say that this knowledge has helped you work with children (humans) with FASD in your clinical training by informing your knowledge of the pathophysiology and developmental trajectory of the disorder, as well as by exposing you to potential cutting-edge treatments that may soon be included as recommendations in neuropsychological reports.

A second key to successful interviewing is to sell the fit between yourself and the site. "Fit" is one of those nebulous concepts that is difficult to define. Broadly speaking, it is the degree to which your career/internship goals align with those of the site. We believe that the best way to demonstrate fit is through preparedness, enthusiasm, and clear communication about the overlap between your aims and objectives and those of the program. When communicating about the high level of fit, we encourage you to reference the names of specific supervisors and clinics. For example, you might say,

> My career goal is to provide cognitive assessment and intervention services to young adults with acquired brain injuries in a rehabilitation setting. This is one of the reasons why I am incredibly excited about the Neuropsychological Services rotation in the Rehabilitation and Recovery Clinic. In particular, Dr. Nguyen's work with transcranial magnetic stimulation

for stroke patients is of great interest to me. I don't have any TMS experience myself, but I have read about it and attended talks on it, and I would love the chance to gain clinical experience in this modality by working with him.

Notice a few important aspects of this quote. First, the applicant specifically delineates the overlap between herself and the program in terms of her future plans and one of the rotations offered at the site. Second, she names the rotation and one of the supervisors who works there. By name-dropping in this manner, you demonstrate to the interviewer that you have done your homework and read up on their program. It shows conscientiousness and interest. Third, this applicant is also up front about the fact that she is interested in transcranial magnetic stimulation (TMS) but does not have prior experience with it. This is an important point. Some graduate students think that, in order to be a good fit with an internship, they should have experience in all of the settings, clinical populations, and tools offered by the site. They mistakenly believe that this is what is meant by "fit." This is not the case. Internship is meant to help budding clinicians round out their training and this necessarily means acquiring new experiences. So, an important part of the fit between the site and the applicant is that the applicant has not already done *everything* offered by that site. We say all of this to ease your concern if you do not have experience in a particular setting (e.g., rehabilitation), population (e.g., stroke), or methodology (e.g., TMS). Don't try to hide the gaps in your training. Lean into them, while championing your areas of strength and experience, and express interest in novel training opportunities offered by the site of interest.

Lastly, any time a site is interviewing you, don't forget that you are also interviewing them. Students with strong applications are highly sought after by internship programs and it is common for faculty and staff to "roll out the red carpet" and attempt to maximize the appeal of their site to competitive applicants. So, if you find yourself overcome with nerves and self-doubt, try to reframe your negative thoughts such that you view yourself as the interviewer and the site as the interviewee. You are there to ascertain the fit between yourself and them, and to determine whether their program offers the opportunities that would meet your needs.

Match Procedure

All APA-accredited programs adhere to an APPIC match process, which is administered by National Matching Services, Inc. (NMS).[59] The match is based on an algorithm that pairs participants with the best possible site based on the preferences of both parties (i.e., applicants rank all potential sites and vice versa). The math underlying the algorithm is complex, but your job is simple. Your sole duty is to

---

[59] Visit APPIC's website to view a number of training resources available to students, including links to the American Psychological Association of Graduate Students (APAGS): https://www.appic.org/Training-Resources/For-Students.

rank sites/tracks in order of your preference. That's it. Don't overthink the situation by altering your ranking based on guesses about where particular programs may rank you. The system is designed for both parties to make individual decisions while remaining blind to the other party's choices.[60]

The match is carried out in two stages: Phase I and Phase II. All registrants participate in Phase I; here, match results are released in late February ("Match Day"). Unmatched participants and program openings can participate in Phase II, which is an accelerated version of the initial application and interviewing process. Phase II results are released in late March. In the unlikely event that you still have not matched in Phase II, there is a Post-Match Vacancy Service (PMVS) that lasts until the Fall. This is a relatively unstructured process that allows students to apply directly to sites that list vacancies on the PMVS website. For a much more detailed account of these processes, visit the APPIC (https://www.appic.org/Internships/AAPI) website and read the *Internships in Psychology* book mentioned earlier.

Some sites have multiple tracks, which are ranked independently. In other words, if there are non-neuropsychology tracks that offer sufficient neuropsychological training (e.g., in geropsychology, rehabilitation, child psychology), then you can apply to those in addition to the more traditional options. For example, imagine that you apply to Internship X, which has two tracks of interest – the neuropsychology track and the geropsychology track. If you are very interested in Internship X, then the top of your rank-ordered list might look like this:

1. Internship X, neuropsychology track
2. Internship Y
3. Internship Z
4. Internship X, geropsychology track

Internship X is on your list twice because the two separate tracks each require an individual ranking.

Before moving on, we have one final piece of advice to share. Because the results of the match are binding, only list sites/tracks where attendance would be preferable to not matching at all. That is, do not apply to a site simply to increase your chances of matching *anywhere*. In other words, if you imagine yourself there and decide that you would prefer to stay at your graduate program for another year rather than completing their internship, then do not include them in your rank-ordered list.

Action steps:

 Follow the application timeline and steps in Table 5.

 Apply to 15–20 internship sites, all of which look to be a good fit for your interests and career goals.

---

[60] For more information about how the match works, including a brief video, go to https://natmatch.com/psychint/algorithm.html.

To prepare for interviews, role play the question-answer session, with a focus on summarizing your dissertation and delineating the fit between yourself and the site.

 Rank your sites based on the order of your preference and no other factors.

Completing Internship

Given that experiences vary dramatically across interns, we can only provide general advice regarding the internship year itself. First, you can anticipate a unique adjustment period during this time. After multiple consecutive years at your graduate institution, it may feel unsteady to be in a new place, with entirely new supervisors, staff, and colleagues. This will also be your first formal 9-5 clinical job in quite some time, if ever, which will be a shift in and of itself. Know that these feelings are normal.

Second, you will likely be provided with more autonomy during internship compared to graduate school. If you are well-versed in all aspects of the neuropsychological evaluation, this may be a nice change of pace; internship will provide you with more flexibility, as well as the ability to simulate the life of a professional neuropsychologist. On the other hand, don't be shy about asking for more supervision in areas where you lack sufficient training. This is a great time to address any gaps in your skills, as you will be expected to have attained a higher level of proficiency by the time you begin fellowship. So, expect to push the boundaries of your comfort level. For example, you may be placed in a fast-paced interdisciplinary environment such as a rehabilitation setting, where you are required to work quickly and efficiently, serving multiple patients per day while consulting with nurses, occupational therapists, and physicians, and checking in frequently with your supervisor. If you have not operated in such an environment in the past, it will likely be a shock to your system initially. Keep in mind that your supervisors will allow for an adjustment period and do what you can to adapt as quickly as you can, while maintaining a high level of quality in everything that you do.

Third, the internship year is all about rotations. A rotation refers to a discrete clinical experience and is analogous to a graduate school practicum. Given that you will have read the brochure and interviewed at the site, you will already have a good sense as to which rotations are available prior to beginning the internship. Then, at some point between match day and your first day of internship, you will be placed into particular rotations for the year. Sites go about the process of rotation selection in a variety of ways. For example, in some situations, you may match to a site, and then go through a second ranking process for rotations. Along with the rankings of the other incoming neuropsychology interns, your rankings are taken into account and the training director decides who goes where, and when. Alternatively, you may be the only neuropsychology intern, so it may be a given that you will complete the

Cognitive Assessment Rotation, and your primary decision may center around the selection of secondary rotations. Overall, this is one of those issues that is heterogeneous across sites, but you will know what you are in for when you browse the program's materials.

Fourth, remember that this is a predominately clinical year, but all sites supplement clinical work with didactics and some also allow for a small amount of protected research time. Didactics will take place in the form of regular seminars, some more general (e.g., ethics, psychotherapy), and others more targeted (e.g., transplant evaluations, neuroanatomy).[61] Research may occur in the context of a small amount of protected time per week (typically four hours – rarely, eight hours). Alternatively, there may not be any protected time and you may be relegated to working on these projects in your spare time, during evenings and weekends.

Action steps:

 Talk to supervisors early in the year about the gaps in your training and how to address them.

 Sign up for rotations that will add up to at least 50% time in neuropsychological work.

 Keep an open mind and work to adapt quickly to new clinical settings.

 Supplement clinical work with didactics and research to the extent possible.

Applying to Postdoctoral Fellowship

Overview

Just when you are beginning to feel settled in your internship position, about two to three months after your start date, the time comes to prepare for postdoctoral fellowship applications. The short turnaround is tough because many people feel as though they just finished the internship application process. Next is the upheaval around adjusting to a new home and job and, before you know it, you are searching for fellowship sites and updating your CV once again. Fortunately, this time around, the process is not as time-intensive as it was during internship applications. For example, most fellowships do not require essays and there is no requirement for tracking each and every clinical hour. However, there are some unique aspects of this process that are important to know.

Postdoctoral fellowship (or "postdoc") represents the culmination of your neuropsychology training. It comprises two years of full-time education and training (although it can be completed on a half-time basis, meaning four years of training).

---

[61] Per the Taxonomy, a student must complete didactic experiences consistent with HC guidelines for knowledge and skill for neuropsychology to be considered a Major Area of Study.

According to the HC guidelines, fellowship is "designed to provide clinical, didactic and academic training to produce an advanced level of competence in the specialty of clinical neuropsychology and to complete the education and training necessary for independent practice in the specialty" (Hannay et al. 1998, p. 4).

Fellowship is the time to truly specialize in the field, which is why it is required for board certification in clinical neuropsychology and why we believe that the title, "neuropsychologist" is reliant upon completion of this capstone training experience. This is akin to a physician being referred to as a "neurologist" only after they complete a neurology fellowship.

The HC guidelines (Hannay et al. 1998, p. 4) make several other specifications regarding fellowship training[62]:

1. The faculty is comprised of a board-certified clinical neuropsychologist and other professional psychologists;
2. Training is provided at a fixed site or on formally affiliated and geographically proximate training sites, with primarily on-site supervision;
3. There is access to clinical services and training programs in medical specialties and allied professions;
4. There are interactions with other residents in medical specialties and allied professions, if not other residents in clinical neuropsychology;
5. Each resident spends significant percentages of time in clinical service, and clinical research, and educational activities, appropriate to the individual resident's training needs.

Note: Regarding #1 above, although this is included in the HC guidelines and we feel that it is important, ABCN does *not* require supervisors to be board certified. However, according to ABCN, supervisors should have demonstrable training and experience comparable to a board-certified neuropsychologist.

We recommend rereading the full HC guidelines document prior to applying to sites to ensure that your training will fulfill all of these requirements.[63] You are at the mercy of the credential review committee as to whether they will allow you to move through the board certification process if you do not strictly adhere to the guidelines. Therefore, we suggest that you either adhere to every detail or that you reach out to the board to inquire about your situation *before applying* to a fellowship.[64] Also, don't be shy – explicitly ask training directors for clarification about whether they adhere to these standards. You have come too far to jeopardize your future certification eligibility.

---

[62] Fellows who have been affected by disruptions to their training as a result of COVID-19 should document all service delivery and supervision activities, and refer to ABCN's memo for more information on the topic: https://theabcn.org/covid-19-postdoctoral-training-memo/.

[63] The document also includes "exit criteria" that are useful to know ahead of time.

[64] You can email the ABCN Credential Review Committee chair at Credentialreview@theabcn.org.

Timeline and Required Materials

Application deadlines are typically in November through January of the fellowship year but we recommend that you begin looking into your options in September in the event that one of your preferred sites has an earlier deadline. As the months go by, stay informed about available fellowship slots because some openings may not be announced until later in the year. In terms of application materials, this process is much less structured than was internship and there is no standard set of requirements. Most sites will request an updated CV, a few sample reports, a cover letter/letter of interest, graduate school transcripts, and three letters of recommendation.[65] Some sites ask for applications to be submitted to the APPIC online platform, while others request that materials be emailed.[66] Regarding the letters of recommendation, by this point, you likely have more than three neuropsychologists who know you well and could write excellent letters, so you may be in the fortunate position of not knowing whom to ask. One solution is to ask people at the same site to coauthor one letter. So, you could ask all of your internship supervisors to write one letter and then ask several mentors from graduate school to write the other two. If you are applying for clinical positions, then it is important that at least one of the letter writers be a board-certified neuropsychologist (Driskell et al. 2020).[67] As always, request letters several months in advance and provide writers with helpful information (e.g., deadlines, list of sites you are applying to, CV).

Research Training Models

Unlike internship, which is primarily a clinical experience, fellowships vary greatly in the emphasis placed on clinical work compared to research. Many postdoc programs in neuropsychology are largely (75-85%) clinical, closely mirroring internship. Others lean toward research experience (50-100%), seeking to train the next generation of academic scholars. We cover the details of clinical fellowships below, so we will focus on research positions here. Typically, research-focused fellowships are grant-funded. A common mechanism for this is the National Institute of Health (NIH) T32 (https://researchtraining.nih.gov/programs/training-grants/t32) and this happens to be the type of program that I (Ryan) attended. For those of you who may

---

[65] These materials will be very similar to those submitted for internship applications (with the addition of internship experience), so please refer back to the internship section for specific advice.

[66] APPIC has several online resources for students applying to fellowship, including a sample spreadsheet to organize programs. See https://www.appic.org/Postdocs.

[67] To add data to this, Driskell et al. (2020) found that 43% of training directors said it is "essential" that letters of recommendation be written by neuropsychologists, 43% said it is "very important," and 14% said it is "somewhat important."

be interested in this option, the experience will depend greatly on (a) the personality and style of the faculty mentor and (b) the research topic. There is typically a high degree of autonomy and unstructured time, although the fellow works in the lab of the advisor and so is expected to focus on that person's research area. A rough trajectory of the T32 experience is as follows:

Year one: publish papers with the mentor and close collaborators, building a publication record together

Year two: continue publishing and apply for a NIH Career Development (K) Award (https://researchtraining.nih.gov/programs/career-development/K01)

Year three (optional): revise and resubmit the K Award if necessary

The K Award is a multi-year grant that launches the trainee's career as a researcher. By the end of the grant, it is expected that the fellow will be well-prepared to write their own large-scale grants (e.g., R01: https://grants.nih.gov/grants/funding/r01.htm) and build a career as a scientist. If you are interested in this type of career, we highly recommend The Grant Application Writer's Workbook (Robertson, Russell, & Morrison, 2020).

In addition to research, many T32 programs will allow the fellow to flexibly allocate their time to other activities as agreed upon by the trainee and mentor. For example, it is possible to incorporate clinical work, teaching, and a variety of professional activities into one's schedule. Indeed, the plasticity of these positions, as opposed to higher degrees of structure in more clinically oriented positions, often allows for greater diversity in training.[68]

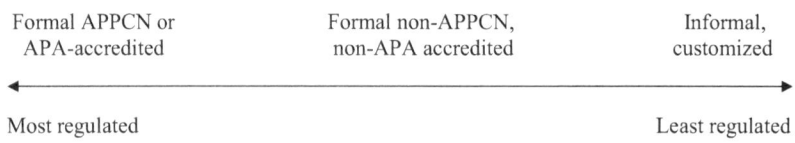

| Formal APPCN or APA-accredited | Formal non-APPCN, non-APA accredited | Informal, customized |

Most regulated   Least regulated

Fig. 6.2 Spectrum of postdoc training programs from most to least regulated/established

---

[68] Completing a research-only fellowship will not make the individual eligible for board certification in clinical neuropsychology. However, even in a T32 program, it is often possible to structure your experiences in a manner that preserves eligibility for board certification in clinical neuropsychology – for example, by negotiating clinical experiences sufficient to meet ABCN guidelines, and by documenting neuropsychological reports and feedback to patients as part of research studies. See Chapter 7 if you are interested in the details.

Clinical Training Models

Although clinical fellowships are more structured than research fellowships, they are less standardized than graduate school/internship, and program accreditation is not mandatory. As such, several clinical models are viable, and we present them in Fig. 6.2. Keep in mind that these are not formalized training frameworks – they are heuristics that we use to conceptualize the available clinical options. On the more regulated, established end of the continuum are Association of Postdoctoral Programs in Clinical Neuropsychology (APPCN) and APA-accredited programs.[69] On the unstructured side of the continuum are customized, "choose your own adventure" fellowships. In the middle lie several additional options.

First, APPCN is a group of postdoc programs that have standardized the training experience offered at each site in order to ensure that they adhere to the HC guidelines. Per their website, APPCN programs "offer the highest quality competency-based residency training in clinical neuropsychology with an emphasis on preparation for future specialty board certification though the American Board of Professional Psychology/American Board of Clinical Neuropsychology (ABPP/ABCN)" (https://appcn.org/who-we-are/our-history-and-mission/). As such, completing an APPCN program essentially guarantees that you will pass the credential review part of a board certification application. Not only that but the neuropsychology boards allow candidates to skip part of the ABCN application if they completed an APPCN fellowship.[70]

Second, as we mentioned, you can still become a board-certified neuropsychologist by attending a non-APPCN, non-APA-accredited program. There are many such programs that offer high-quality training and have a long history of producing board-certified neuropsychologists. Typically, these programs explicitly state that they adhere to the HC guidelines in their online materials, and we suggest that you ask for this to be included in the official offer letter before accepting a position. Additionally, we recommend confirming that prior graduates of the program have successfully become boarded – this is a good indicator that you will pass the credential review as well.

Third, furthest down the continuum are people who stitch together fellowships that provide the requisite training for board certification. The most common approach here is to find an open position at a local group/private practice and to create a 2-year supervised clinical position under the guidance of the staff neuropsychologist(s). During that time, didactics, scholarly work, and interdisciplinary interactions can be acquired either at a nearby university or at the practice

---

[69] APA accreditation is technically the highest degree of regulation because APA conducts a detailed review of these sites. APPCN membership has an internal approval system. Also note that APA accredits programs as a whole (i.e., all psychology postdocs) and specialty programs in neuropsychology.

[70] ABCN also lets you skip items on the application form if you completed an APA-accredited specialty program in clinical neuropsychology.

itself. For example, Dr. Danielle Bello, currently a board-certified neuropsychologist through ABPP/ABCN, created her postdoc in a group practice. She fulfilled the didactics requirement by completing NAN's online neuroanatomy course, creating a weekly seminar at her practice, attending neurology rounds at a local university medical center, and joining a neuropsychology seminar series through a colleague's training program.[71] For the research requirement, she analyzed preexisting clinical data from the practice and presented the findings at multiple conferences. The appeal to this approach is that it allows the fellow great flexibility in finding an open clinical practice in a desirable geographic location. However, the onus is on you to ensure that your fellowship adheres to the HC guidelines.[72] Also, there are a few details to consider regarding the ABCN approval process[73]:

1. ABCN has explicitly stated that the fellowship must "reflect a structured and sequenced set of clinical and didactic experiences, provide on-site supervision of all clinical cases, and put the learning needs of the candidate ahead of the operational needs of the program."

2. Similar to #2, ABCN has specific criteria that must be met if fellowship training is completed in more than one setting: "Applicants must demonstrate that the fellowship training across the respective sites was *structured, sequenced and well-integrated*, and that the trainee's role in both settings was as a post-doctoral resident/trainee (rather than other roles, such as staff psychologist or research coordinator)." There are some documentation requirements in this situation, so refer to ABCN's site for details.

3. Regarding the clinical-research split, according to ABCN, "Training in neuropsychology-related activities must comprise a minimum of 50% of the resident's time and may include supervised clinical activities done as part of research. However, at least 50% of the clinical neuropsychological services must involve integrative neuropsychological evaluation (i.e., services that include a component of clinical service delivery, including integration of results in a written report to patients, research participants, or physicians) under the supervision of a clinical neuropsychologist."

4. The site must have a sufficient number of neuropsychology-specific didactics. As part of the ABCN credential review, you will demonstrate to the board that you have received didactics in eight topic areas: basic neurosciences, clinical neurology, clinical neuropsychological assessment, functional neuroanatomy,

---

[71] For these reasons, it is important to consider the location of your self-created fellowship.

[72] For example, we suggest that people who are considering the customized approach be extra mindful of the need to interact with other medical/allied health residents, which might not be available in a typical group practice setting.

[73] Each of the quotes comes from https://theabcn.org/credential-review-frequently-asked-questions/. Other ABCN webpages to peruse include https://theabcn.org/becoming-certified/ and https://abpp.org/Applicant-Information/Specialty-Boards/Clinical-Neuropsychology/Application,-Specialty-Specific-Fees.aspx.

Table 6.7 Resources for locating neuropsychology fellowships

Resource	Website
APPIC Directory	https://membership.appic.org/directory/search
Universal Psychology Postdoctoral Directory	https://www.appic.org/Postdocs/Universal-Psychology-Postdoctoral-Directory-UPPD
APPCN	https://appcn.org/member-programs/
Society for Clinical Neuropsychology (SCN) Directory	https://scn40.org/training-directory/
APPIC email list	https://www.appic.org/Postdocs
APA-Accredited Postdoctoral Programs	https://accreditation.apa.org/about

neuropathology, psychopathology, psychological assessment, and psychological intervention. Although you can list didactics and less structured educational activities (e.g., case conferences, brain cutting conferences, specialty rounds) completed in graduate school and internship, it is expected that a substantial proportion of your didactic training will come from fellowship (no percentage is specified). According to ABCN's website: "While didactics may be listed from different eras of one's training, some form of didactic training must be integrated into the post-doctoral program, and it would be expected that postdoctoral training would be represented in most topic areas."

Of course, all fellowships must abide by these principles, but it is more likely that the board will scrutinize customized positions more carefully than established fellowships. As such, our general advice is to aim for an established program with a track record of producing board-certified neuropsychologists if possible. If you do choose to create your own fellowship, we recommend that you discuss this with your current supervisors and reach out to the board before committing to the path.[74] Also, we encourage you to get written affirmation from your prospective postdoctoral supervisor that the training will meet the criteria listed in the HC guidelines.[75]

Selecting Sites

You know the drill. The process of researching and selecting potential fellowship sites mirrors the process for internship.[76] In Table 6.7, we list several resources to help you get started. Similar to internship applications, we suggest relying heavily

---

[74] Again, you can email the ABCN Credential Review Committee chair (Credentialreview@the-abcn.org) if questions remain after you read through the HC guidelines and ABCN website materials.

[75] Also, consider logging hours and reviewing the ABCN application at https://abpp.org/Applicant-Information/Specialty-Boards/Clinical-Neuropsychology/Document-Library.aspx.

[76] To keep organized, we recommend you use an excel spreadsheet. Templates are available at www.NavNeuro.com/book and https://www.appic.org/Postdocs.

on the APPIC Directory and searching for fellowships with "neuropsychology" in the description. Finally, keep an eye out for fellowship postings on the major neuropsychology listservs (e.g., Npsych) as well.

In terms of the number of sites to apply to, your list will be significantly smaller than it was for internship. Although there is no gold standard recommendation here and there is a great deal of variability in how applicants approach this process, we suggest that most people consider applying to approximately 8–12 sites.[77] To arrive at that number, you can consult with your advisor to first brainstorm/investigate and create a pool of potential options and then work on narrowing the field. In this context, we provide specific advice below.

First, as we mentioned above, an option for many students is to apply to postdoc at the same institution that houses their internship. Advantages to staying put center around the fact that the fellowship will be a known entity and there will be less disruption and more continuity in the training experience. Specifically, (a) the staff and facilities will be familiar, (b) relationships with faculty supervisors will already be established, (c) ongoing research projects can be maintained, and (d) geographic relocation will not be required. This is not to say that completing internship and postdoc at the same institution is the right choice for everyone. Some excellent internship programs do not offer fellowships and some trainees benefit from the diversity in training and supervision that comes from an additional transition.[78] Moreover, there is no guarantee that a fellowship program at a particular site will accept current interns when they are also reviewing applications from external candidates. The bottom line here is to keep your options open and consider your own site (if possible), as well as other programs.

Second, let's talk about geography. For graduate school and internship we recommended that you cast as broad of a net as possible by applying to sites nationwide. For fellowship, there are two general options: (a) prioritize the best training possible, regardless of where the program is geographically located, or (b) prioritize the best training possible within a select geographic radius. There are myriad reasons to prefer one region over another; in the case of postdoc, people often select the location in which they want to apply to jobs and settle down after the two years are over. Proponents of option (b) posit that it is important to begin networking in the general region where they will be applying to jobs, and that it is not uncommon for fellows to be hired on as staff/faculty at their fellowship site. On the other hand, proponents of option (a) argue that the best possible training will create the most competitive job candidate regardless of location or institution. Of course, there is no right or wrong answer and we recommend that each person consider all factors relevant to their own situation and make the decision that is best for them.

---

[77] Applicants who successfully matched to a fellowship site in 2020 applied to an average of 6.7 programs (National Matching Services Inc 2020). Note that this number does not include applications submitted to non-match programs.

[78] If you end up staying at the same program for fellowship, consider asking questions such as, "Will the didactics be presented at a higher level?" and "How will postdoc training be different from internship?" (e.g., will you have access to new clinical populations or supervisors?).

Third, it is important that you begin working primarily with the clinical population(s) you want to serve as an independent practitioner. For example, if you wish to work with children with neuromedical disorders, then pediatric neuropsychology in a hospital setting is likely the right place for you. In contrast, if you plan to seek out a job in a rehabilitation clinic, then apply to fellowships that consist of experiences with interdisciplinary teams and post-acute traumatic brain injury and stroke patients. Remember that postdoc is your chance to specialize in your area of interest. You already have the requisite generalist training under your belt, so do not be afraid to go for depth during this time.

Interviewing

Fellowship interviews, while important, will likely be far less structured and evaluative than were internship interviews. By this point in your training, and based on your CV, it is assumed that you possess the requisite skills of a budding neuropsychologist. Consequently, the process is mostly about the fit between your goals/needs and those of the site. As we mentioned above, this is the capstone training experience that will solidify your identity as a neuropsychologist, so use the interview as a way to ensure that the site will meet your needs and that the personality and style of your supervisor-to-be fit with your own.

Although some programs require that candidates travel for an on-site interview, most are traditionally held at the February INS meeting, especially for APPCN programs.[79] This is obviously less expensive and more convenient overall. However, there are unique challenges to this approach. Namely, the INS conference hotel typically does not have enough space to allow for individual private rooms. Instead, one large room is reserved and interviews take place at adjacent roundtables. Interviewees describe this as a challenging cognitive task, given that they are forced to ignore many other conversations going on around them during their own session. What's more, there can be a speed dating vibe to the process, where a candidate interviews with one site at one table, and then moves a few tables down for the next interview. Thus, it can be hard to remain focused on the conversation at hand. Overall, there is a lot to juggle in these interviews, but many people have successfully navigated this process, and you can, too.

In addition to formal interviews, we recommend that you seek out current or recent fellows in the program and ask to talk with them about their experiences. Conversations such as these are often chock full of helpful advice and tips. Finally, for those of you who are interested in additional interviewing advice, we recommend our NavNeuro episode on the topic (www.NavNeuro.com/08).

It is important to talk to trainees outside of the formal interview process… you can get a lot of great info for the price of a beer.

– Kathleen Fuchs, PhD, ABPP-CN

---

[79] Due to the COVID 19 pandemic, in-person interviews may be less common in 2021 and beyond.

Match vs Non-match Whereas all accredited internships are required to adhere to a matching process, this is not the case for fellowships. As of the time of writing, all APPCN member programs and some non-APPCN sites participate in the match, but many other programs do not. Believe it or not, the lack of consensus around a post-doctoral match is one of the most contentious topics in the field of neuropsychology today.[80] Unfortunately, this dual system is dissatisfying to many trainees (Towns et al. 2018), making it less than ideal. Nevertheless, there are currently both "match" and "non-match" sites, so you will need to navigate this quagmire if you apply to both types of sites. Our advice is to simply apply to your top-rated programs without consideration of whether they are match or non-match sites. In terms of how to manage offers, we provide some recommendations in the next section.

Managing Offers

If you applied to sites outside of the match, you may find yourself in the difficult position of receiving multiple offers, each with its own timetable. Although receiving postdoc offers is obviously a positive experience, it can be a difficult process to navigate. For example, imagine that you receive a time-limited offer from your #3 site, but you haven't even interviewed at your #1 site yet and your #2 site is part of the match. What should you do? There is no right answer here. Your decision will depend on a number of idiosyncratic factors such as the actual distance between the sites in terms of your preference (i.e., how much better is #1 than #3?), how long the initial site allows you to hold the offer (typically only a day or two), whether you have backup options (e.g., are sites #4 and #5 acceptable?), and your risk toler-ance.[81] We know people who accepted an offer from their lower-ranked site as well as people who took the gamble, so to speak, and turned down the initial offer in the hopes that their #1 site would eventually accept them as a fellow. Either way, we recommend that you inform your highest-ranked site when you receive an offer. Sometimes they will provide information that will sway your decision (e.g., APPCN programs are allowed to inform an applicant about whether or not the person would be expected to match with them).[82] Once you have accepted a position, it is expected that you immediately inform all other sites of your decision. If you are participating in the match, you should know that the results are binding; so, once you submit your rank order list, you cannot accept a position from a non-match site until after the match (if you do not match).

---

[80] For an in-depth discussion of the pros and cons of the match process, see Bodin and Grote 2016a; Nelson et al. 2016; Bodin and Grote 2016b.

[81] As you can tell, it is essential that you think through your rank order before you begin receiving offers.

[82] APPCN includes advice for how to handle this situation: https://appcn.org/applicants-faq/.

Action steps:	
	Familiarize yourself with postdoc requirements outlined by the HC guidelines and ABPP/ABCN.
	Determine the type of fellowship that is most consistent with your career goals: research-oriented or clinically focused.
	Create an initial list of 12–15 possible sites and then narrow it down to 8–12.
	Read and prepare for the match/non-match process ahead of time and, if necessary, consider a plan on how to handle multiple offers.

Self-Care and Stress Management

We have seen that graduate school is a challenging endeavor, with numerous stressors and challenges. Consequently, we want to say a few words about the importance of self-care and stress management.

One day in graduate school when I (John) was feeling particularly stressed, I mentioned my stress level to a professor who responded, "No one looks back at grad school fondly; just push through it." While we understand the sentiment, we do not agree that graduate school must be a harrowing experience. Importantly, there are factors under your control that you can alter in order to dramatically improve your graduate school experience, and we provided specific recommendations in the previous sections to help you avoid many common pitfalls. Here, we will provide general stress management advice. First, try to determine whether or not a particular stressor is within your control. This is important because psychological research suggests that, for stressors within our control, the best response is to use a problem-focused coping technique (try to solve the issue and improve your situation). However, for those stressors that are unavoidable and unchangeable, such as a difficult course or a challenging living situation, emotion-focused coping is more helpful (accept the situation and focus on healthy, mood-enhancing activities; Carver and Connor-Smith 2010).

Prior to coping with particular situations, or "putting out fires," so to speak, there are multiple methods to partially inoculate yourself against general stress and worry. These techniques tend to improve both mental and physical health, no matter the person's environment or life stage. Here are some of our favorites:

- After familiarizing yourself with the path to your goal and formulating a plan, focus your attention on taking one step at a time.
- Keep your thoughts and materials organized and prepare in advance for upcoming assignments/responsibilities.
- Practice self-confidence (e.g., during your first therapy experience) and remind yourself that you were selected to be in your graduate program.
- Foster an environment of collaboration – not competition – with your classmates. Remember, they will be your neuropsychology colleagues for the foreseeable future.
- *Always* prioritize sleep (except in class… unless you can master the advanced technique of sleeping sitting up, eyes-open).

- Engage socially.
- Exercise and eat a varied, nutritious diet.
- Practice mindfulness, meditation, yoga, deep breathing exercises, and/or other restorative practices.
- Spend time outside daily.
- Maintain a routine and daily structure.
- Reframe negative thoughts to be positive or neutral.
- Consider therapy/counseling with a professional if you feel depressed and/or highly anxious.
- Know that it is normal to feel overwhelmed at times.
- Know that it is also normal to experience "imposter syndrome."
- Frequently remind yourself of the reasons why you chose to pursue this career path. Reread Ch. 2 if necessary.

Setbacks

Successfully navigating graduate school necessitates the completion of a variety of major tasks, and setbacks are possible at each and every point along the way. For example, you may not have received the ideal practicum placement, you may not have passed an important examination (e.g., the comprehensive exam), you may not have passed your dissertation proposal, you may not have matched to internship, and/or you may not have been accepted to a postdoc. With so many steps, it is hard to imagine anyone who has not encountered multiple setbacks. If you find yourself in the unenviable position of encountering one of these roadblocks, know that *you are not alone.* Many outstanding neuropsychologists have histories of grappling with the stress, frustration, and self-doubt that comes along with facing a major disappointment during training. While we cannot provide you with individualized advice beyond what is already available in this chapter, we can provide a few general tips. First, if you are struggling emotionally with the bad news, take a few days off (if possible) and do something fun and restorative. Second, after you are feeling a bit better, try to pinpoint areas of growth and improvement so that you can experience success the next time around. Third, talk to trusted advisors and friends/colleagues about your situation and solicit advice and support.

Conclusion

Graduate school is a time of tremendous professional and academic growth. You have come a very long way from your first day as a bright-eyed, bushy-tailed graduate student to being a doctorate-level clinical psychologist, well-prepared for a postdoctoral fellowship. By this time, you will have tackled advanced coursework, challenging research projects, and a few thousand hours of clinical work. The finish line to becoming a neuropsychologist is now right around the bend.

References

Alden, A., Van Tuyl, L., Chow, J., Davis, C., Del Rio, R., Peruzzi, N., & Rodolfa, E. (2000). *1994–1999 Intern Applicant Practicum Hours: An Exploratory Investigation.* Poster presented at the 108th Annual Convention of the American Psychological Association, Washington, DC.

Association of Psychology Postdoctoral and Internship Centers. (2020). *APPIC match statistics.* https://www.appic.org/Internships/Match/Match-Statistics

Association of State and Provincial Psychology Boards. (2009). *ASPPB guidelines on practicum experience for licensure.* https://cdn.ymaws.com/asppb.site-ym.com/resource/resmgr/guidelines/final_prac_guidelines_1_31_0.pdf

Bieliauskas, L. A., & Steinberg, B. A. (2005). The evolution of training in clinical neuropsychology: From hodgepodge to Houston. In G. J. Lamberty, J. C. Courtney, & R. L. Heilbronner (Eds.), *The practice of clinical neuropsychology* (pp. 17–30). Lisse: Swets and Zeitlinger.

Bodin, D., & Grote, C. L. (2016a). Commentary: The postdoctoral residency match in clinical neuropsychology. *The Clinical Neuropsychologist, 30*(5), 641–650.

Bodin, D., & Grote, C. L. (2016b). Rebuttal to Nelson et al. 'Response to Bodin and Grote regarding postdoctoral recruitment in clinical neuropsychology'. *The Clinical Neuropsychologist, 30*(5), 660–663.

Bodin, D., Butts, A. M., & Grote, C. L. (2016). Postdoctoral training in clinical neuropsychology in America: How did we get here and where do recent applicants suggest we go next? *The Clinical Neuropsychologist, 30*(8), 1371–1379.

Carver, C. S., & Connor-Smith, J. (2010). Personality and coping. *Annual Review of Psychology, 61*, 679–704.

Donders, J. (Ed.). (2016). *Neuropsychological report writing.* New York: The Guilford Press.

Driskell, L. D., Del Bene, V. A., & Sperling, S. A. (2020). How to become a competitive neuropsychology intern and postdoc applicant. *Know Neuropsychology.* https://knowneuropsych.org/how-to-be-a-competitive-fellow-intern/

Duff, K., O'Bryant, S. E., Westervelt, H. J., Sweet, J. J., Reynolds, C. R., Van Gorp, W. G., et al. (2009). On becoming a peer reviewer for a neuropsychology journal. *Archives of Clinical Neuropsychology, 24*(3), 201–207.

Dweck, C. S. (2007). *Mindset: The new psychology of success.* Random House Publishing Group.

Frank, E., & Levenson, J. C. (2010). *Interpersonal psychotherapy.* American Psychological Association.

Gauthier, B., Dupont, C., Gosselin, N., & de Guise, E. (2020). Neuropsychology supervision: A survey of practices in Quebec and a cross-cultural comparison. *The Clinical Neuropsychologist.* Advance online publication. https://doi.org/10.1080/13854046.2020.1732467

Gregory, E., Soderman, M., Ward, C., Beukelman, D. R., & Hux, K. (2006). AAC menu interface: Effectiveness of active versus passive learning to master abbreviation-expansion codes. *Augmentative and Alternative Communication, 22*(2), 77–84.

Hannay, H. J., Bieliauskas, L. A., Crosson, B. A., Hammeke, T. A., Hamsher, K. deS., & Koffler, S. P. (1998). Proceedings of the Houston conference on specialty education and training in clinical neuropsychology. *Archives of Clinical Neuropsychology, 13*(2), 157–158.

Hayes, S. C., Strosahl, K. D., & Wilson, K. G. (2012). *Acceptance and commitment therapy: The process and practice of mindful change* (2nd ed.). New York: The Guilford Press.

Heffelfinger, A. K., Janecek, J. K., Johnson, A., Miller, L. E., Nelson, A., & Pulsipher, D. T. (2020). Competency-based assessment in clinical neuropsychology at the post-doctoral level: Stages, milestones, and benchmarks as proposed by an APPCN work group. *The Clinical Neuropsychologist*, 1–17.

Hessen, E., Hokkanen, L., Ponsford, J., van Zandvoort, M., Watts, A., Evans, J., & Haaland, K. Y. (2018). Core competencies in clinical neuropsychology training across the world. *The Clinical Neuropsychologist, 32*(4), 642–656.

INS-Division 40 Task Force. (1987). Reports of the INS-Division 40 task force on education, accreditation, and credentialing. *The Clinical Neuropsychologist, 1*, 29–34.

Keilin, G. (2018). *2018 APPIC match: Survey of internship applicants part 1: Summary of survey results.* https://www.appic.org/Internships/Match/Match-Statistics/Applicant-Survey-2018-Part-1

Magana, A. J., Vieira, C., & Boutin, M. (2017). Characterizing engineering learners' preferences for active and passive learning methods. *IEEE Transactions on Education, 61*(1), 46–54.

Michel, N., Cater, J. J., III, & Varela, O. (2009). Active versus passive teaching styles: An empirical study of student learning outcomes. *Human Resource Development Quarterly, 20*(4), 397–418.

Miller, W. R., & Rollnick, S. R. (2013). *Motivational interviewing: Helping people change* (3rd ed.). New York: The Guilford Press.

National Matching Services Inc. (2020). *APPCN resident matching program: Summary results of the match for positions beginning in 2020.* https://natmatch.com/appcnmat/stats/2020stats.pdf

Nelson, A. P., Roper, B. L., Slomine, B. S., Morrison, C., Greher, M. R., Janusz, J., et al. (2015). Official position of the American Academy of Clinical Neuropsychology (AACN): Guidelines for practicum training in clinical neuropsychology. *The Clinical Neuropsychologist, 29*(7), 879–904.

Nelson, A., Bilder, R. M., O'Connor, M., Brandt, J., Weintraub, S., & Bauer, R. M. (2016). Response to Bodin and Grote regarding postdoctoral recruitment in clinical neuropsychology. *The Clinical Neuropsychologist, 30*(5), 651–659.

O'Connell, B. (2012). *Solution-focused therapy* (3rd ed.). Sage.

Rabin, L. A., Paolillo, E., & Barr, W. B. (2016). Stability in test-usage practices of clinical neuropsychologists in the United States and Canada over a 10-year period: A follow-up survey of INS and NAN members. *Archives of Clinical Neuropsychology, 31*(3), 206–230.

Ritchie, D., Odland, A. P., Ritchie, A. S., & Mittenberg, W. (2012). Selection criteria for internships in clinical neuropsychology. *The Clinical Neuropsychologist, 26*(8), 1245–1254.

Robertson, J. D., Russell, S. W., & Morrison, D. C. (2020). *The grant application writer's workbook: National Institute of Health version.* Grant Central, LLC.

Schwent Shultz, L. A., Pedersen, H. A., Roper, B. L., & Rey-Casserly, C. (2014). Supervision in neuropsychological assessment: A survey of training, practices, and perspectives of supervisors. *The Clinical Neuropsychologist, 28*(6), 907–925.

Smith, G., & CNS. (2019). Education and training in clinical neuropsychology: Recent developments and documents from the clinical neuropsychology synarchy. *Archives of Clinical Neuropsychology, 34*(3), 418–431.

Sperling, S. A., Cimino, C. R., Stricker, N. H., Heffelfinger, A. K., Gess, J. L., Osborn, K. E., & Roper, B. L. (2017). Taxonomy for education and training in clinical neuropsychology: Past, present, and future. *The Clinical Neuropsychologist, 31*(5), 817–828.

Stucky, K. J., Bush, S., & Donders, J. (2010). Providing effective supervision in clinical neuropsychology. *The Clinical Neuropsychologist, 24*(5), 737–758.

Svinicki, M. D., & McKeachie, W. J. (Eds.). (2014). *McKeachie's teaching tips: Strategies, research, and theory for college and university teachers* (14th ed.). Wadsworth: Cengage Learning.

Sweet, J. J., Perry, W., Ruff, R. M., Shear, P. K., & Breting, L. M. G. (2012). The Inter-Organizational Summit on Education and Training (ISET) 2010 survey on the influence of the Houston Conference training guidelines. *Archives of Clinical Neuropsychology, 27*(7), 796–812.

Sweet, J. J., Klipfel, K. M., Nelson, N. W., & Moberg, P. J. (2020a). Professional practices, beliefs, and incomes of postdoctoral trainees: The AACN, NAN, SCN 2020 practice and 'Salary Survey'. Archives of Clinical Neuropsychology.

Sweet, J. J., Klipfel, K. M., Nelson, N. W., & Moberg, P. J. (2020b). Professional practices, beliefs, and incomes of US neuropsychologists: The AACN, NAN, SCN 2020 practice and "salary survey". *The Clinical Neuropsychologist, 1–74*.

Towns, S. J., Hahn-Ketter, A. E., Halpern, J., & Block, C. K. (2018). Trainee perspectives on postdoctoral recruitment in clinical neuropsychology: Reflections on commentaries by Bodin and Grote (2016) and Nelson et al. (2016). *The Clinical Neuropsychologist, 32*(1), 10–15.

Williams-Nickelson, C., Prinstein, M. J., & Keilin, W. G. (2019). *Internships in psychology: The APAGS workbook for writing successful applications and finding the right fit* (4th ed.). American Psychological Association.

Wright, J. H., Brown, G. K., Thase, M. E., & Basco, M. R. (2017). *Learning cognitive-behavior therapy: An illustrated guide* (2nd ed.). American Psychiatric Association Publishing.

Chapter 7

Advanced Training and Practice: Postdoctoral Fellowship and Beyond

Learn something new every day, do good work, and have fun doing it.

– Steve Correia, PhD, ABPP-CN

Graduate school is a monumental hurdle, and making it through the gauntlet is a praiseworthy accomplishment. But your training is not quite finished. In order to become a clinical neuropsychologist, you are required to complete a 2-year neuropsychology fellowship, pursue licensure as a psychologist, and (we hope) obtain board certification in neuropsychology. In this chapter, we will walk you through each of these steps and touch upon securing employment, independent practice, managing student loans, and advocacy work. We also provide a brief overview of the re-specialization process for those of you who have not taken a traditional path into the field.

Postdoctoral Fellowship

Postdoc is the time to truly specialize in neuropsychology.[1] By now, your generalist training is complete, and you can immerse yourself in your area of deepest interest. Many neuropsychologists say that fellowship was the most enjoyable and rewarding

---

[1] As we spelled out in Ch. 6, it is crucial for neuropsychology fellowships to adhere to the Houston Conference (HC) guidelines and American Board of Clinical Neuropsychology (ABCN) requirements (see the *Board certification* section, below). For this reason, all of our advice in this chapter is consistent with these training models. We strongly recommend that you educate yourself and use these guidelines as a compass while navigating the postdoc experience.

© Springer Nature Switzerland AG 2021
J. A. Bellone, R. Van Patten, *Becoming a Neuropsychologist*,
https://doi.org/10.1007/978-3-030-63174-1_7

stage of their training. It allows for greater autonomy than did graduate school while still providing support and guidance as needed, and it prepares the trainee to transition into fully independent practice. Fellowship is also your last chance to fill in any gaps in knowledge and skills while still in a formal training role. As such, it is important to make your wishes and expectations known early on by sitting down with your advisor and creating an individualized education plan that includes the professional experiences you need and desire in your last years as a trainee.[2]

Postdoc training varies widely across institutions and people. For example, our respective experiences differed significantly, despite the fact that we both fit squarely under the neuropsychology umbrella. I (Ryan) completed a 2-year research fellowship in the Geriatric Mental Health T32 at UCSD, where I focused on neuropsychological research, supplemented with clinical work, teaching, and professional development. The program is open to scientists of many ilks, including neuroscientists, psychiatrists, data scientists, and exercise physiologists, along with psychologists and neuropsychologists. In addition to mandatory research productivity, I contacted faculty at UCSD and nearby universities to offer my services as a clinician, teacher, and consultant. As a result, I remain eligible to apply for board certification in clinical neuropsychology. Although the goal of the fellowship is to train academic scientists to apply for National Institute of Health (NIH) Career Development (K) Awards, there was great flexibility in the overall training experience, and I enjoyed frequent interdisciplinary collaborations and an intellectually rich work environment.

On the other hand, I (John) completed a formal 2-year clinical neuropsychology fellowship at Brown University. I was housed at the Providence VA where I regularly provided services to Veterans with a variety of neurocognitive issues. In this setting, I gained depth in serving Veteran populations. However, it also afforded me significant breadth in terms of setting and testing approach. For example, I provided services to Veterans from different eras (e.g., WWII, Vietnam, the conflicts in the Middle East) and engaged in teleneuropsychological evaluations, home-based testing, community outreach through a self-created "Lifestyle and Brain Health" education and consultation service, neuropsychology didactics, neuroradiology rounds, neuropathological autopsies (i.e., "brain cuttings"), and multiple research projects (20% of my time was protected for research). By the end of fellowship, I felt very prepared for independent clinical practice.

$$* * * *$$

One helpful way to conceptualize the postdoc experience is as the transition phase between student who is supervised and colleague who supervises and consults. Technically, fellows are still trainees (i.e., they are not autonomous clinicians, and they are not the PIs of major grants), but in practice they often function more as early career faculty than graduate students. This becomes even truer as the trainee progresses through the fellowship program. In other words, we are advocates of a

---

[2]Consider using the competencies listed in Smith and CNS 2019, to guide that conversation.

titrated approach to training, where the degree of supervision tapers off and the degree of independence simultaneously expands over the course of the experience. Specifically, the first year of fellowship will still involve oversight and the smoothing out of areas of inexperience, with gradual increases in autonomy. The second year then involves a slow removal of all training wheels. In the clinical realm, we believe that the transition from fellow to full-fledged neuropsychologist is smoother and more natural if the trainee is provided with the opportunity to act as a fully independent professional, albeit with the buffer of a behind-the-scenes supervisor who can step in if necessary. And we are not alone in this belief.[3]

> The most valuable experience I had as a trainee was about two months prior to the end of my postdoc, when my supervisor, John Beetar, stopped supervising me. He had me conduct clinical evaluations and write neuropsych reports and then he provided guidance as needed. And it was terrifying, but it gave me a safety net... the opportunity to be a grown up psychologist before I was actually out on my own. So I think that if trainees can work with supervisors to allow for more independence on the back end of postdoc, it will make them stronger clinicians later on.
>
> – Kira Armstrong, PhD, ABPP-CN

In practice, this clinical safety net approach may involve the trainee independently performing the records review, clinical interview, testing, scoring, and report writing and then discussing the case with the supervisor as needed before finalizing the report and providing feedback. We agree with Dr. Armstrong that several months of this type of training serve as excellent preparation for the next stage. If you are fortunate enough to have this experience, we encourage you to adopt the mindset of an independent clinician. Imagine that it is your license number and signature at the bottom of the report and that you are responsible for all aspects of the case. This is a major responsibility that we all take very seriously, and it can lead to anxiety in freshly minted clinical neuropsychologists. Although some level of anxiety and worry are natural and can motivate attention to detail, we know from the famous Yerkes-Dodson curve (Yerkes and Dodson 1908) that too much can be paralyzing and can curtail performance. So, lean into the new responsibilities, do your due diligence for each and every case, and round out your skills as a neuropsychologist, all in preparation for the start of a successful career.

Although postdoc programs vary wildly and we will not presume to be able to provide individually tailored advice, we do have a few points of consideration, which we provide first for research-focused fellows and second for clinically oriented fellows.

Research fellowships:

- Publish! Peer-reviewed journal articles are the currency of the academic world, so make this a priority.

---

[3] Of course, a licensed professional should always provide the degree of clinical oversight required by state law.

- Submit applications for grants and awards. These can include smaller supplemental awards such as the Society for Clinical Neuropsychology (SCN) Early Career Pilot Study Award (https://scn40.org/sac/) and the International Neuropsychological Society (INS) Nelson Butters Award (https://www.the-ins.org/about-ins/ins-awards/nelson-butters-award-recipients/), as well as large grants such as the NIH K Award (https://researchtraining.nih.gov/programs/career-development/K01) and the VA Career Development Award (CDA) (https://www.research.va.gov/funding/cdp.cfm).

- Narrow your focus. This is a good time to select a niche research area if you have not done so already.

- Maximize efficiency without sacrificing quality.

  - This means collaborating with interdisciplinary teams where tasks are delegated accordingly. For example, in the context of writing papers, undergraduates can perform literature reviews and handle references and formatting, colleagues can share co-first authorship, statisticians can consult on complex analyses, and experts can provide relevant guidance and resources. You do not need to shoulder each and every manuscript task by yourself – be a team player and reap the benefits.
  - This also means improving your own personal approach to grant writing and paper composition. Are you minimizing distractions and making the most of your time? Everyone has their own style, and it is important that you find the routine that works well for you. I (Ryan) spend a great deal of time searching the literature, reading, taking notes, and making outlines before I ever write a single word. Using this approach, I have found that the task of actually writing a paper is relatively smooth and straightforward, and my output has increased accordingly.

Clinical fellowships:

- Adopt your supervisor's report template, but also develop your own style. Try to absorb the good and leave the not-so-good as you begin forming your own approach to important high-level tasks such as case conceptualization, clinical interviewing, test selection, report writing, and feedback.

- To the extent possible (and ethical), test out different models of, and approaches to, clinical care. For example, try providing extended feedback that includes several sessions of cognitive training before referring patients to other providers. Try out different ways of explaining difficult and complex topics in feedback sessions (see www.NavNeuro.com/29). Experiment with different neuropsychological tests (see www.NavNeuro.com/30).

- Seek out new experiences. For example, teleneuropsychology is likely to remain an important approach to clinical evaluations for the foreseeable future (see www.NavNeuro.com/41), so consider asking your supervisor if you can participate in remote assessment and intervention.

- Maximize efficiency without sacrificing quality. This means taking advantage of report templates (https://iopc.online/report-writing), utilizing dictation software, and cutting unnecessary time from your clinical routine. Also, be on the lookout for a standardized, computerized history form and other exciting technological developments (www.NavNeuro.com/33).

- Remain engaged in research. Many clinical fellowships in neuropsychology are 10–25% research, and this is an important part of the capstone experience. Lean into these opportunities and align your scientific work with your clinical interests.

Action steps

 Look up and adhere to the Houston Conference (HC) Guidelines and ABCN requirements for postdoc training in neuropsychology.

 Sit down with your advisor and plan for the remainder of your fellowship training.

 Request a "hands-off" approach to supervision for the last several months of your fellowship.

 Work to maximize efficiency without sacrificing quality.

Psychology Licensure

Licenses are designed to protect the public. They signify that the person has achieved a minimal level of competency in order to deliver safe and effective services.

– Joel Kamper, PhD, ABPP-CN

Up until now, you have been providing clinical care under the license of a supervisor. In order to begin seeing patients as an independent clinical neuropsychologist, you will first be required to obtain a psychology license in a particular state or province.[4,5] This is an important point: the requirements for licensure in psychology vary by jurisdiction. Consequently, we cannot provide specific advice. Each individual applicant must look up and then follow the requirements for the area in which they wish to practice. Of course, many people do not know where they will settle down, and this process can become complicated when someone decides to move

---

[4] If you seek a purely academic (research/teaching) career, then licensure is not necessary. However, we recommend that all neuropsychologists interested in clinical research also become licensed, even if provision of clinical services is only a small fraction of their professional time. This both a) diversifies a neuropsychologist's professional skills portfolio and b) improves their scientific acumen by putting them in touch with clinical populations.

[5] You will almost always be a licensed psychologist practicing neuropsychology, not a "licensed neuropsychologist."

from one region to another. For now, we will refer you to a full NavNeuro episode on the topic (www.NavNeuro.com/42) and advise that you begin the process of applying for licensure as soon as you are eligible. Typically, a state or province will require a particular number of graduate and postdoctoral clinical hours prior to submission of the application, and many trainees will become eligible during their second year of fellowship, so this is a common time to begin focusing on this task.

There are several steps to becoming licensed, and we will touch upon these steps throughout the remainder of this section. Again, keep in mind that much of the following information varies by jurisdiction.

Step #1: Determine the state(s)/province(s) in which you will seek licensure and look up/follow the requirements for licensure in those jurisdictions.

First, know this: you are specializing in neuropsychology and are required to complete a 2-year fellowship as part of the HC guidelines (most specialties in clinical psychology do not require a 2-year postdoc). This experience, together with an APA-accredited graduate program and high quality/quantity of clinical hours, should prepare you to meet licensure criteria. Still, be diligent and ensure that you are on the correct path. To do so, consider your individual situation. Do you know exactly where you want to live and work in the future? If you answered "yes," then your situation is simpler and more straightforward. Learn the requirements for your jurisdiction and adhere to them. On the other hand, if you are like many psychological trainees, your future home is not yet known, and your situation is more complex. If you can narrow down the candidate jurisdictions to 2–3, then we recommend checking the requirements of each state/province and following them. If you cannot narrow down the candidate jurisdictions, it is obviously untenable for you to look up the specifications of dozens of different states/provinces. One strategy here is to adhere to a set of very strict criteria (e.g., Massachusetts, California), knowing that this will likely cover you for states with less stringent criteria.[6]

Under certain circumstances in the United States, it is possible to practice across state lines without being licensed in both states. Indeed, reciprocity among states is becoming increasingly common, thanks in large part to the Psychology Interjurisdictional Compact (PSYPACT). If you are licensed in one of the PSYPACT states, you can request permission to practice telepsychology/teleneuropsychology and/or deliver in-person services in another PSYPACT state.[7] This process is much easier (and cheaper) than becoming licensed in the second state. In today's world,

---

[6] Here is a list of state, provincial, and territorial agencies responsible for licensure of psychologists throughout the United States and Canada: https://www.asppb.net/page/BdContactNewPG.

[7] Check if your state is part of PSYPACT: https://psypact.org/page/psypactmap.

with telehealth becoming more and more common, there are good reasons for many neuropsychologists to consider applying for these reciprocity arrangements. That being said, some circumstances do call for obtaining licensure in multiple jurisdictions. If this applies to your circumstances, you may benefit from "banking" your credentials and materials in order to ease the application process. Two organizations currently offer this paid service – National Register, https://www.nationalregister.org/apply/credentialing-requirements/, and ASPPB https://www.asppb.net/page/TheBank.[8]

Step #2: Once eligible, collect the required materials and submit your application for licensure.

Each state/province has its own application form and set of required materials. Materials may include academic transcripts, Examination for Professional Practice in Psychology (EPPP) scores, reference letters, and verification of supervised experience.[9] States vary in whether applicants are required to submit these documents before or after taking the EPPP (see below).

Tip: If needed, request signatures before leaving a site. For example, if you know that your internship supervisor will be required to complete a verification of experience form for licensure, then get it signed before you move away at the end of the year. This will save you time and energy later on.

Step #3: Register for the EPPP and begin studying.

Once your application is approved, the state will typically grant you permission to take the EPPP,[10] which is developed and owned by the Association of State and Provincial Psychology Boards (ASPPB). A passing score on the EPPP is required for a psychology license in all jurisdictions, so you will be taking it regardless of the state in which you choose to practice. There are now two parts to this computerized exam: Part 1 (knowledge) and Part 2 (skills).[11]

---

[8] The ASPPB Credentials Bank has partnered with ABPP. Visit https://abpp.org/Applicant-Information/5-Types-of-applications/Early-Entry.aspx for details.

[9] Note that some states require that you register your anticipated postdoc experience and verify supervision *prior to* the start of fellowship, so check the rules of the state in which you plan on becoming licensed before starting your training.

[10] Some jurisdictions have waiting periods before an applicant is allowed to register to take the EPPP. This can cause unfortunate delays, and some applicants choose to register for the EPPP in a different, more lenient state/province than the one in which they live so that they can take the test and get this step out of the way sooner rather than later.

[11] Visit https://www.asppb.net for guidance on how to register for the exam and additional exam information.

Part 1 (*Knowledge*) consists of 225 multiple-choice questions covering large content areas within psychology (e.g., assessment and diagnosis, professional issues, lifespan development, social psychology). A score of 500 is considered passing in most jurisdictions at the doctoral level, and this score translates to having answered 70% of items correctly. All jurisdictions require Part 1.

Part 2 (*Skills*) is designed to test the examinee's ability to apply their knowledge to clinical situations. There are 170 questions.[12] Part 2 is a recent addition to the EPPP and, at the time of this writing, is still being rolled out in particular jurisdictions. In other words, you may or may not be required to complete Part 2. Neither of us has taken Part 2 of the exam, so the remaining advice in this section pertains to Part 1. If you are required to take Part 2 as well, there will be plenty of study materials available through ASPPB and PsychPrep.

People differ in their approach to preparing for Part 1 of the exam. Your graduate school training will have prepared you to answer many of the questions correctly; indeed, the test is designed to assess the same content areas as were covered in graduate classes. Still, studying will increase the probability that you will pass on your first attempt. Although examinees are allowed multiple attempts to pass the test, it is expensive and time-consuming to take, and not passing may delay your licensure and potential employment. With this in mind, our recommendation is to begin studying early (4–5 months in advance) and begin at a slow pace, with a gradual increase in frequency as the exam approaches and heavy studying for 1–2 weeks prior to the exam. Research shows that spaced practice/distributed learning such as we are recommending leads to more efficient memory consolidation than does cramming (Cepeda et al. 2006; Dunlosky and Rawson 2015). Moreover, there is an enormous amount of content to cover for the EPPP, so cramming is even less effective here than it is when taking an exam in college or graduate school. Consequently, we caution against procrastination.

So if you agree with us about when and how much to study, you may be wondering *what* to study. There are several formal programs that offer resources for a fee. We both used and enjoyed PsychPrep (https://psychprep.com), but there are several other options as well. Typically, these programs include written study guides, practice tests, and audio files, all covering the relevant topic areas. If you are interested in more in-depth training, some programs also offer workshops and individualized expert feedback. We both found it very useful to take in the material from several different modalities (i.e., written study guides, audio files, and online tests), and we recommend this approach.[13]

---

[12] For sample items and other resources, see https://www.asppb.net/page/EPPPPart2-Skills.

[13] In addition to online tests through third parties, ASPPB provides access to retired questions from prior exams for a small fee: https://www.asppb.net/page/Practiceexinfo.

Step #4: Take any additional exams, if needed.

In addition to the EPPP, some states/provinces require an additional exam, which can be administered in either written or oral format and typically covers ethical and legal considerations pertaining to mental health in that particular jurisdiction (i.e., "jurisprudence exam"). As always, check the requirements in the jurisdiction(s) in which you are applying for licensure.

Step #5: Submit your final documents and pay your licensure fee.

Action steps

 If possible, determine the jurisdiction(s) in which you will practice and then look up and adhere to their licensure requirements.

 Begin studying for the EPPP early and space out your studying initially, with gradual increases in time spent studying as the exam approaches.

Securing Employment

Finally! You are no longer a trainee. After so many years of study, it is now time to seek employment as an independent psychologist and clinical neuropsychologist. We will keep our guidance broad because the process of securing employment is as variable as the number of settings and geographic locations available (see Ch. 3). Once you decide on the *where* (i.e., setting and region), you can begin actively searching for openings. Generally speaking, we recommend that you begin this process in earnest about 7–10 months before your fellowship ends.

Here are some methods for identifying openings:

- Rely heavily on your network. Tell your current and past supervisors and colleagues that you will be "on the market" soon and ask them to notify you if they hear about any openings. Don't be shy about this – we recommend a low threshold for sending these emails. Additionally, if you are interested in a job at your current fellowship site, talk with your supervisor about the possibility of applying to an open position and/or creating an on-site position.
- Where appropriate, send cold emails to department chairs and other decision-makers at facilities that are of interest to you.
- For government positions (e.g., VAs), sign up to be notified of openings at https://www.usajobs.gov. Many large healthcare providers have similar notification lists.
- Sign up for and monitor neuropsychology listservs (https://scn40.org/scn-list-servs/; www.neurolist.com) where job openings are frequently posted.
- Check the notification boards at neuropsychology conferences.

- Periodically check webpages of the major neuropsychological organizations (e.g., https://theaacn.org/view-jobs/; https://nanonline.org/jobbank/; https://www.the-ins.org/job-postings/).
- Search major professional network and career development websites (e.g., Indeed, Glassdoor, Lensa, LinkedIn).
- Simply Google "neuropsychologist job in [location]."
- Consider joining a local state, provincial, or territorial psychological association (SPTA), and networking with psychologists and neuropsychologists in your area. This is a great way to meet people who will be able to help you find a job in your desired region. Visit https://www.apaservices.org/practice/advocacy/state/spta for more information.

Regarding the job interview, see Ch. 6 and www.NavNeuro.com/08 for general advice. Similarly, if you receive multiple job offers, see Ch. 6 for tips on how to think through this decision. Although these resources focus on internship and fellowship, much of the advice applies to your current situation as well.

Once you have secured an offer, do not be afraid to negotiate for better compensation and benefits. We do not learn negotiation skills as part of our graduate or fellowship programs, so we have little, if any, formal training in this area. However, these skills help us in both salary discussions and in daily interpersonal interactions, so the downstream benefits are worth the upfront cost. For a primer, consider reading the book, *Negotiation Genius: How to Overcome Obstacles and Achieve Brilliant Results at the Bargaining Table and Beyond*, by Malhotra and Bazerman (2007). If you do not currently have the bandwidth for an entire book, Deepak Malhotra (2014) also wrote a helpful article entitled, *15 Rules for Negotiating a Job Offer*, published in *Harvard Business Review*.[14]

Specific to neuropsychology, we suggest that you leverage data from the most recent salary survey (Sweet et al. 2020a) to justify your requested compensation amount, adjusting the number based on your individual experiences, work setting, and the cost of living in your area. We understand that the negotiation process can feel awkward and uncomfortable for psychologists who are trained to communicate in a very different manner. However, keep in mind that most employers expect a negotiation to take place, and there is nothing to lose by making reasonable requests or simply asking whether there is any "wiggle room." So remind yourself that you are a well-trained professional who deserves fair compensation for the work that you do. Read up on resources for negotiation such as those we provided above. And, if you feel anxious, use one of the many effective techniques to ameliorate social/performance anxiety and do not let your fear get in the way of self-advocacy.

---

[14] There are also resources specific to women such as the book *Women Don't Ask: The High Cost of Avoiding Negotiation – and Positive Strategies for Change* by Babcock and Laschever; SCN's Women in Neuropsychology Subcommittee (https://scn40.org/piac-win/); and NAN's Women in Leadership Committee (https://www.nanonline.org/NAN/_AboutNAN/Committee_Pages/Women_in_Leadership.aspx).

The initial job negotiation will likely impact your salary for years to come, so this process is well worth your mindful attention and effort.

Importantly, there are many relevant job-related factors in addition to salary. Below, we list several of the variables that we encourage you to inquire about and (if possible) negotiate for. Even settings that have a set pay schedule and benefits package (e.g., VA hospitals) may be flexible with respect to miscellaneous factors such as office space, support staff, compressed work week, etc. Note that some of what we discuss below will be setting specific (e.g., fee-for-service, research/academic).

- Referral sources and patients

  - Who are the main referral sources?
  - What are the most common referral questions and cases?
  - Is the patient population demographically diverse?
  - How steady/reliable is the stream of referrals?
  - Are you required to see a minimum number of patients per week?

- Compensation/promotion

  - What is the compensation model? (e.g., salary, fee-for-service)

    - How is your value measured? (e.g., revenue generation, number of patients assessed, number of publications)
    - Will you receive a raise when you become board certified?

  - If fee-for-service[15,16]:

    - What is the percentage split? (i.e., what is the percentage of the total reimbursed amount that you keep after the company/owner takes their share? In our experience, this ranges from 50–65%.)
    - How often are claims not fully reimbursed? Does the company offer any protection against lost compensation? (e.g., can they guarantee that you will receive a certain amount of money per case?)
    - Is there any protection in the event that the referral stream ebbs?
    - Which insurance providers does the group accept and what is the average number of cash-pay patients?
    - What is the typical take-home amount per case?

---

[15] This applies primarily to clinicians in private/group practice.

[16] If you are considering a job in private practice or taking on a role with administrative responsibilities, the book *The Business of Neuropsychology* (2010) by Mark Barisa is a helpful resource.

- If academic[17]:

  - What does the academic review/promotion process look like? That is, what is required to move from assistant professor to associate professor to full professor? (Examples include publications, grants, teaching, and service.)
  - Is tenure offered? If so, how does one achieve tenure?
  - Is the position funded primarily through grants or through department funds?
  - Are there opportunities for sabbaticals? If so, what is offered?

- Benefits package

  - Does the company offer medical insurance, liability coverage, paid time off, retirement contributions, professional development funds, and/or coverage of licensure and board fees?

- Job duties

  - Are you responsible for marketing and bringing in new clinical referral sources/patients?
  - Are you responsible for administrative tasks? (e.g., scheduling, payment receipt)
  - Do you have access to a psychometrist to aid in testing/scoring? If so, does this affect your salary/percentage split?
  - Do you have access to research assistants to aid with data collection/management?
  - Are there students (e.g., practicum students, interns) and/or fellows to supervise?
  - What percentage of your time will be devoted to teaching/research/clinical work?
  - Will you receive an academic affiliation with employment, and what are your responsibilities to maintain that affiliation? (e.g., teaching)

- Miscellaneous

  - Is the institution willing to buy you new equipment? (tests, software, record forms, laptops, etc.)
  - How much autonomy do you have over the assessment battery and report format?
  - What safety precautions are in place at the facility?
  - What does IT support look like?
  - To what journal subscriptions will you have free access?
  - To what statistical software will you have free access?

---

[17] Also consider reading the book, *The Professor Is In: The Essential Guide to Turning Your Ph.D. into a Job* (2015) by Karen Kelsky, PhD.

Another crucial job-related factor that is not typically negotiable but is neverthe-less critical to well-being is the length of the daily commute. There is a growing literature showing negative correlations between commute length and health out-comes (e.g., Hansson et al. 2011; Künn-Nelen 2016), which is not surprising given the time demands, monetary cost, and sedentary nature of this task. Of course, some commutes (e.g., quietly working on a train or riding a bicycle) are more enjoyable and productive than others (e.g., anxiously negotiating rush hour traffic). Consequently, we encourage you to carefully consider the length and quality of your daily commute and then weigh it heavily when considering your job options.

****

Securing job offers and navigating this process of negotiation and decision-mak-ing can be stressful, as it is a major life decision. We want you to be thoughtful, mindful, and decisive. We also want you to free yourself from unnecessary pressure and burden because you will have other opportunities down the road.

Your first job doesn't have to be your forever job.

– Christine Koterba, PhD, ABPP-CN

Action steps

 To find job openings, make it known to everyone in your professional network that you are on the job market. Also monitor listservs, notification boards, and relevant websites.

 After securing offers, read up on negotiation tactics and then implement them as appropriate. Use the Sweet et al. (2020a) salary survey to justify your requested compensation amount.

 Inquire about and negotiate for important fringe benefits.

 Take steps to reduce daily commute time.

Independent Neuropsychology

No matter how independently you functioned during fellowship, it will always be a shift to move from trainee to self-sufficient practitioner. There is a great deal of responsibility that comes with serving as the Principal Investigator (PI) on major grants, supervising and teaching students, and signing off on clinical reports. It may feel strange to submit a clinical note or a research paper without asking an advisor to review the document first. However, it is entirely normal to doubt your compe-tence at times and to wonder whether or not you made a mistake. We have a few pieces of advice if you find yourself plagued by doubt and/or imposter syndrome:

1. A modest degree of concern about the validity of your work can be beneficial – it can motivate you to increase your conscientiousness and attention to detail. So, inasmuch as these feelings lead you to read a few more journal manuscripts and/ or spend more time preparing for a new patient, lean into them.

2. Too much self-doubt is certainly not healthy and attenuates performance (think of it as related to stress/anxiety and then invoke the Yerkes-Dodson law). In this case, remember that you successfully completed years of rigorous academic training and cultivate confidence in your professional abilities. If this does not help and you are still struggling, talk to trusted friends and colleagues.

3. Regardless of your degree of confidence/self-doubt, keep in touch with mentors and colleagues. Do not feel shy about consulting with others, especially when you are faced with challenging clinical cases, complex statistical analyses, or any other difficult problem. Neuropsychology involves lifelong learning and collaboration, and we encourage you to always turn to knowledgeable peers and seasoned professionals when in doubt.

4. Take advantage of early career resources. For example, SCN has an Early Career Neuropsychologist Committee (ECNPC) that provides guidance to individuals such as yourself (https://scn40.org/about-join-ecnpc/). APA has a similar committee, the Committee on Early Career Psychologists (CECP, https://www.apa.org/careers/early-career/).

5. Blaze your own trail and find your own style. You are no longer following the preordained path from college through fellowship. You will now have much more room for creativity and a greater degree of decision-making ability in determining your career path as a neuropsychologist. Take advantage of this.

****

Independent neuropsychology is a universe unto itself, and volumes could be written on this topic alone. But the purpose of this book is on how to *become* a neuropsychologist, not how to *be* a neuropsychologist, so we will focus on two important aspects of early professional development and allow your training and job experience to do the rest.

Continuing Education

Always be curious and consider yourself a student and a learner.

– Julie Hook, PhD, MBA, ABPP-CN

Becoming an independent neuropsychologist does not signify the end of new learning. Even 4 years of college, 6 years of graduate school, and 2 years of fellowship are not nearly enough to cover *all* of the information and skills that could be relevant to a neuropsychologist. Moreover, the knowledge turnover in medicine and psychology is rapid, so even if you did learn everything there is to know about neuropsychology by the end of fellowship, it would still be necessary to continue updating your internal models to account for new evidence.

> I am on PubMed every day of my life. With my students, if I can communicate one thing, it's that we are always, always learning.
>
> – Karen Postal, PhD, ABPP-CN

Clearly, there are good reasons to continue learning throughout your career. But even if you aren't convinced, the licensing boards are. To our knowledge, every jurisdiction requires that psychologists complete a certain number of continuing education (CE) credits per renewal period in order to maintain their license. And both teaching/supervision and research necessarily involve continued consumption of the scientific literature as well. Below, we provide options for remaining relevant and up to date in neuropsychology, and many of the options listed come with formal CE credits.

- Read neuropsychology journals.[18]

  - *Journal of the International Neuropsychological Society* (https://www.the-ins.org/education/online-continuing-education/)
  - *Neuropsychology*
  - *Neuropsychologia*
  - *Archives of Clinical Neuropsychology*
  - *The Clinical Neuropsychologist* (https://aacncontinuingeducation.org/)
  - *Neuropsychology Review*
  - *Child Neuropsychology*
  - *Journal of Pediatric Neuropsychology*
  - *Journal of Clinical and Experimental Neuropsychology*
  - *Applied Neuropsychology*

- Listen to the NavNeuro podcast (www.NavNeuro.com/INS).
- Complete formal online didactics.

  - International Neuropsychological Society (INS, https://www.the-ins.org/education/ce/)
  - National Academy of Neuropsychology (NAN, www.nanonline.org → "Continuing Education")
  - American Academy of Clinical Neuropsychology (AACN, https://aacncontinuingeducation.org/)

---

[18] Many scientific journals in related fields are also likely to be relevant to neuropsychologists. Consider journals in neurology, neuroscience, psychology, psychiatry, assessment, statistics, pediatrics, geriatrics, rehabilitation, and others.

- Engage in neuropsychology-focused webinars and lecture series (e.g., https://knowneuropsych.org).
- Monitor listservs and engage with colleagues.
- Attend conferences (i.e., INS, NAN, AACN, APA) and complete CE workshops.
- Read and re-read test manuals to reduce "administrator drift."
- Frequently consult with colleagues.

Preventing Burnout

As we discussed in Ch. 4, neuropsychologists often wear many hats. Together with increasing expectations in terms of patient care and scientific productivity, this can be a recipe for burnout. If you find yourself in this position, we have included some advice in Ch. 4 and 6. We will also refer you to APA resources here: https://www.apa.org/monitor/2018/02/ce-corner. The take-home point is, no matter how much you enjoy your work, your health and well-being are more important, so align your behavior with your values.

Managing Student Loans

If you are like most doctoral graduates, you are saddled with significant student loan debt. Specifically, a recent APA survey reported a median graduate debt of $98,000 for early career professionals (Doran et al. 2016). We know that this can feel burdensome and stressful. We do not have a simple solution to this massive problem, but we have some general information that we hope will prove useful. Keep in mind that we are not financial advisors and are not offering personalized recommendations.

There are two main options when it comes to eliminating student loan debt. Option 1 is to pay off the full amount at your own pace. This offers you the most flexibility because there are no restrictions regarding work setting or payoff timeline, aside from a minimum payment amount. People who choose this option often refinance the loans to obtain the lowest possible interest rate. However, interest continues to accrue, so paying off the loan quicker saves money.

Option 2 is to commit to a loan forgiveness or repayment program. There are a number of different types of programs, and the details are complex, so an in-depth discussion is outside the scope of this book.[19] One of the more popular options is the

---

[19] There are many excellent resources available for both clinicians and researchers: https://www.apa.org/apags/resources/affording-repaying; https://www.ed.gov; https://www.lrp.nih.gov; https://studentaid.gov/manage-loans/repayment/plans; https://theaacn.org/financial-resources/.

Public Service Loan Forgiveness (PSLF) Program (https://studentaid.gov/manage-loans/forgiveness-cancellation/public-service). To qualify for the PSLF, you must be employed by a US jurisdiction or not-for-profit organization, work full time, have direct loans, agree to an income-driven repayment plan, and make a certain number of qualifying payments (typically 10 years' worth). The upside to this type of program is that, over the course of the payment period, you could pay significantly less than the total loan amount. However, there are several downsides. First, the stipulation regarding working for federal or specific organizations limits your employment options and flexibility. Second, your loans and repayment plan must meet particular criteria. Of course, it is crucial that the details be followed scrupulously in order to avoid jeopardizing your standing in the program.

If you meet criteria for option 2, then the decision about whether to choose option 1 or 2 comes down to both the math and your own personal values. For example, even if it entails paying slightly more money by choosing option 1, some people prefer not to have this type of debt hanging over their head for years or decades. Of course, given the stakes and the level of complexity, we recommend that you seek professional guidance early on (preferably while still on fellowship) regarding this important decision. To this end, you could contact your loan servicer, reach out to a third party that specializes in student loans (e.g., Student Loan Planner, www.studentloanplanner.com), and/or consult with a local certified financial planner (CFP).[20]

There are a variety of books, podcasts, and blogs on personal finance management. There are also several helpful expense tracking platforms and budgeting programs. We will share a few of our favorites because we think they are very helpful for living well and avoiding/managing debt:

Books

- *Personal Finance For Dummies*, by Eric Tyson (2019)
- *ChooseFI: Your Blueprint to Financial Independence*, by Chris Mamula, Brad Barrett, and Jonathan Mendonsa
- *The Total Money Makeover: A Proven Plan for Financial Fitness*, by Dave Ramsey

Podcasts

- ChooseFI (start with episode 100)
- The White Coat Investor

Blogs

- Mr. Money Mustache (https://www.mrmoneymustache.com)
- Mad Fientist (https://www.madfientist.com)

---

[20] As an aside, whenever you interact with a CFP, we recommend that you ask how they are paid and whether they have a *fiduciary* relationship with you (meaning that they have to act in your best interest). We suggest going with someone who is a fiduciary and who you pay an hourly rate for their time (as opposed to being paid on commission or through an assets under management model).

Budgeting and expense tracking

- YNAB (https://www.youneedabudget.com)
- Personal Capital (https://www.personalcapital.com)

Advocacy and Outreach

> Early organizational involvement is really important. So many of my mentors were involved
> in national organizations and I looked up to them and saw the contributions they were able
> to make and how they were able to impact the field.
>
> – Munro Cullum, PhD, ABPP-CN

An often overlooked but important aspect of early career neuropsychology is advocacy and outreach. In other words, in addition to our core responsibilities as teachers, clinicians, and researchers, how can we use our skills and platforms to further neuropsychology and help the general public? There are a number of ways to do so, and this work not only benefits the field and wider communities of people but also benefits you as the neuropsychologist. Below are some options to consider:

- Join committees in professional organizations (i.e., INS, AACN, NAN, APA). For example, go to www.nanonline.org → About NAN → NAN Committees – Get Involved.
- Join SPTAs and work at the jurisdiction level in conjunction with APA to advance psychology as a science and profession. For example, see https://illinoispsychology.org/ or https://wspapsych.org/. Also see www.NavNeuro.com/43 for more information.
- Volunteer to speak about topics such as neuropsychological assessment and brain health to your local community and to patient advocacy groups.
- Engage in mentorship in neuropsychology. See the following resources for more information: https://theaacn.org/student-mentorship-program/ and https://scn40.org/piac-ema/.
- Establish an online presence (e.g., through a blog) to provide evidence-based information and answers regarding brain health.

Board Certification

We have made it explicit several times throughout this book that board certification is the highest clinical credential available in our field, and we believe that it is in the best interest of each clinical neuropsychologist to do what it takes to earn these stripes. Moreover, we support the mission of the American Academy of Clinical Neuropsychology (AACN) – promoting board certification as the standard for competence in the practice of clinical neuropsychology (https://theaacn.org). The reasons for endorsing this achievement are numerous, including benefits to the profession, patients, and the practitioners themselves (Armstrong et al. 2019; Cox 2010). As of 2017, neuropsychologists made up only 9% of licensed psychologists, yet they made up 26% of board-certified psychologists (https://www.apa.org/monitor/2017/09/datapoint), thereby underscoring the importance that neuropsychologists assign to board certification. With that in mind, the purpose of this section is to lay out the steps to reach this goal.

There are currently two boards that certify psychologists in general clinical neuropsychology: the American Board of Clinical Neuropsychology (ABCN) and the American Board of Professional Neuropsychology (ABN). A nuanced explanation of the histories and differences between the boards is outside the scope of this book. However, we will focus our attention on board certification through ABCN for the following reasons:

1. ABCN is a member of, and is regulated by, the American Board of Professional Psychology (ABPP).
2. ABPP is the oldest peer-reviewed board for psychology and regulates most of the professional psychology specialties recognized by APA (AACN Practice Guidelines 2007).
3. ABPP is psychology's analog to the American Board of Medical Specialties (the predominant medical board).
4. ABPP has certified the most clinical neuropsychologists (nearly five times as many), and nearly all neuropsychology trainees who responded to recent surveys said they intend to pursue board certification through ABPP/ABCN (Sweet et al. 2020b; Whiteside et al. 2016).

You should know that the requirements for certification are very similar between the boards, so by following the criteria for ABPP (and ABCN, see below), you will likely also meet criteria for ABN.[21]

Note: The American Board of Clinical Neuropsychology (ABCN) is one of 15 examining boards under ABPP and is responsible for the board certification examination process in clinical neuropsychology. AACN is the membership organization for clinicians who have attained board certification through ABCN.

* * * *

Given that board certification is such an important aspect of our field, you may be wondering when you can begin this process. Importantly, you will want to begin thinking about the ABPP/ABCN eligibility criteria while in graduate school because the prerequisites cover the entirety of your neuropsychological training. The primary eligibility requirements for the application are a) completion of a 2-year

---

[21] For more information regarding ABN, visit https://abn-board.com and http://aabnonline.com.

fellowship in neuropsychology and b) acquisition of state/provincial licensure in psychology (but see below for more details). As such, our advice is to align your training with these requirements and begin the application as soon as possible, ideally a few months after postdoc.[22] Candidates have 7 years from the date of credential approval to complete all required steps (known as the "candidacy period"), and the sooner you tackle the process, the more time you will have to reap the rewards of your hard work. It can also be helpful to view the board certification process as an extension of the postdoc experience – the training mindset is still present, and the knowledge base is as fresh as it will ever be. However, it is never too late to pursue this credential.

If you have completed, or will complete, fellowship at an Association of Postdoctoral Programs in Clinical Neuropsychology (APPCN) program (see Ch. 6) in good standing, then you are virtually guaranteed that it will meet standards for board certification. If you are not in an APPCN program, then the onus is on you to ensure that your fellowship adheres to the HC guidelines (Hannay et al. 1998) and ABPP/ABCN requirements.[23,24] We will not cover all of this content in detail here, but we will provide a couple quick tips.

As you embark on the path to board certification, an excellent primer is the book *Board Certification in Clinical Neuropsychology: A Guide to Becoming ABPP/ ABCN Certified Without Losing Your Sanity*, by Dr. Kira Armstrong et al. (2019). Relatedly, we discussed the purpose and process of certification through ABPP/ ABCN with Dr. Armstrong in a NavNeuro episode (www.NavNeuro.com/28). Additional resources include ABCN's website (https://theabcn.org/resources/) and their mentorship program, where they pair the applicant with a board-certified advisor who provides guidance through the process (https://theabcn.org/mentorship-program/).[25] We list and describe the primary steps to board certification and offer some advice/guidance below.

---

[22] If you completed a portion of your fellowship during the COVID-19 pandemic and want to know more about how this will impact your career, see the following website: https://theabcn.org/covid-19-postdoctoral-training-memo/.

[23] Here are some relevant webpages to be aware of: https://abpp.org/Applicant-Information/Specialty-Boards/Clinical-Neuropsychology/Application,-Specialty-Specific-Fees.aspx; https://theabcn.org/credential-review-frequently-asked-questions/; https://theabcn.org/resources/.

[24] Also see the recent publications by the Clinical Neuropsychology Synarchy (Smith and CNS 2019) and Hessen et al. (2018), which describe competencies needed for clinical neuropsychologists.

[25] Also, see https://theaacn.org/relevance-2050-webinar-series/.

Application and Credential Review

Fortunately, the application process is relatively straightforward. Once you are licensed in psychology and have completed fellowship, go to https://abpp.org and begin the process.[26] You will fill out your information and enclose a CV, transcript, and documentation of internship completion. ABPP also requires two letters of recommendation, preferably from board-certified neuropsychologists.

Once the application is submitted, ABPP will first review your general credentials. At this stage, your doctoral training will be vetted in order to ensure that you graduated from a psychology program that met certain requirements (e.g., accreditation; see *Re-specializing in Neuropsychology*, below). ABPP will also verify that you are licensed as a psychologist in a jurisdiction in the United States, its territories, or Canada. If the reviewers deem your credentials to be satisfactory, they will forward the application to ABCN for the specialty (neuropsychology) review. ABCN will focus its review on your fellowship experiences (both clinical and didactic), ensuring that the program was consistent with the HC guidelines (see Ch. 6 for details).[27]

Either ABPP or ABCN may ask you questions and/or request more detailed information. If you followed the HC guidelines and our advice in this book, then you will easily sail through the credential review process. On the other hand, if you have taken a nontraditional path into neuropsychology, or if you have a gap in your training, then see our section below regarding re-specialization.

Written Exam

Once your credentials are accepted, you will be cleared to take the written exam, which consists of 125 computer-administered multiple-choice questions covering general psychology (including statistics and methodology), general clinical psychology, general psychopathology/neuropathology, brain-behavior relationships, and the practice of clinical neuropsychology (ABCN Candidate's Manual 2020). As always, allow yourself plenty of time to prepare for the exam (see our EPPP advice, above). Fortunately, the content of this exam is more targeted than the EPPP, but do not underestimate the breadth and depth of material covered.

---

[26] There is also an "early entry" option for individuals who are not yet licensed. This option allows trainees to begin the application process for a reduced fee. Find out more at https://www.abpp.org/Applicant-Information/5-Types-of-applications/Early-Entry.aspx

[27] There are different requirements for candidates who completed the doctoral degree before 2005 (https://theabcn.org/becoming-certified/). Also, Canadians are offered a bit more flexibility in terms of fellowship training (https://theabcn.org/canadian-guidelines/).

Everyone has their own method for studying, but here are a few options that are specific to this exam:

- Join a study group (https://brainaacn.org/join-a-written-exam-study-group/).
  - Craft a study plan and schedule with your group.
  - Test your study partner(s) on key terms/concepts.
- Read and re-read the book, *Clinical Neuropsychology Study Guide and Board Review* (2020), edited by Stucky, Kirkwood, and Donders.
- Supplement the Stucky et al. book with relevant journal articles, as well as the following books[28]:
  - *Neuroanatomy Through Clinical Cases*, *2nd edition* (2010), by Blumenfeld
  - *Fundamentals of Human Neuropsychology*, *7th edition* (2015), by Kolb and Whishaw
  - *Neuropsychological Assessment*, *5th edition* (2012), by Lezak, Howieson, Bigler, and Tranel
- Utilize the Be Ready for ABPP in Neuropsychology (BRAIN; https://brainaacn. org) resources. It is not necessary to read every study note or take every mock exam – cherry pick what is helpful to you and use it.
- Create mnemonics, associations, and acronyms for difficult-to-remember syndromes/anatomy. For example, when asked about Sturge-Weber Syndrome, think "S = seizures" and "W = wine-like stain on the face."
- Use spaced repetition/distributed practice rather than cramming.

Practice Samples

After passing the written exam, you will be eligible to submit two neuropsychological evaluations as practice samples. These are reports for cases that you completed independently (it is acceptable for a psychometrist to have administered and scored the tests). The cases "should differ sufficiently to demonstrate a range of clinical knowledge and assessment skill, and should demonstrate clearly that the candidate practices clinical neuropsychology at the specialist level of competence" (ABCN Candidate's Manual 2020, p. 12). In addition to the clinical report, you will include a score summary sheet and all raw data (e.g., record forms, score printouts). It is essential that you adequately de-identify all of these materials (triple check to be sure). These practice samples will be reviewed by three ABCN specialists. If deemed acceptable, you will be allowed to advance to the oral examination.

---

[28] These are just a few of the many useful neuropsychology-related books.

There are a variety of strategies for selecting cases. One approach is to begin saving potentially appropriate cases as soon as you start seeing patients independently in order to provide you with multiple options from which to choose. If you do this, then we recommend that you keep a spreadsheet or list of the clinical reports with relevant details (e.g., age, diagnosis, etiology). We also recommend that you select cases that sit squarely within your clinical wheelhouse. In other words, this is not the time to comb through your records looking for patients with rare conditions, multiple comorbidities, and complex, atypical histories. You want to be able to discuss all aspects of the case and the relevant literature, so go with what you know.

Oral Exam

Once the practice samples are deemed acceptable, you are allowed to move on to the final step in the process: the oral exam. This is a three-part, in-person test that has historically been held in Chicago. Each component lasts 45–55 min. The three components, in no particular order, are as follows:

- A set of questions pertaining to the two practice samples submitted by the applicant. The examiner may ask questions about any aspect of the case or may ask the applicant to explain their work based on the scientific literature and professional standards.
- A fact-finding exercise in which the examiner presents the applicant with information pertaining to a neuropsychological case (not one of their own) and the applicant works through the case conceptualization in a one-on-one conversation with the examiner. The applicant obtains the details of the case through questions, similar to a typical clinical interview, requests the cognitive and medical data, and provides their impressions (e.g., diagnosis, etiology, prognosis) and recommendations. The applicant is allowed to choose either an adult or pediatric case.
- An examination of ethics and professional practice, in which the examiner presents a brief vignette containing ethical/professional issues. The applicant must identify the issues, explain the underlying rationale for the ethical principles involved, and describe corrective actions that they would consider taking to resolve the problem and/or how they would have behaved differently in that situation. Following the vignette discussion, the examiner asks the applicant to describe their clinical practice, professional involvement, and continuing education activities. The examiner may ask questions about different professional issues or ethical dilemmas that the applicant has faced.

For each portion of the exam, it is important for you as the applicant to vocalize your thought process so that your examiner can assess your ability to think through the information. Even if your final impressions are incorrect, you can still pass the exam if your method for arriving at conclusions is sound. At the end of the oral exam, you will be dismissed, and your examiners will collectively reach a decision about whether or not you have passed.

Here are a few pointers for preparing for each component of the oral exam:

Practice Sample Exercise

- Read and re-read your reports. These are your clinical cases and you are expected to be the expert on them.
- Brainstorm for questions that the examiner may ask you and then consider how you would respond to each question.
- Read the literature on all aspects of the case (e.g., medications, differential diagnoses, psychometric test properties, selection of normative data).
- Ask trusted colleagues to vet your practice samples and generate questions/ potential areas of concern.

Fact-Finding

- Participate in mock fact-finding cases, both within your study group and with someone who is already board certified. If possible, also hold a mock session with a board-certified clinician who you do not know well in order to simulate the context of the real exam.
- Study common clinical syndromes so that you are familiar with the clinical presentation, disease course, functional neuroanatomy, and typical cognitive profile for those syndromes. Referring back to the *Clinical Neuropsychology Study Guide and Board Review* book is helpful in this regard.
- Read the book, *The Neuropsychology Fact-Finding Casebook: A Training Resource* (2017), by Drs. Kirk Stucky and Shane Bush. These cases are great to use for mock fact-finding exercises.
- Several podcasts have episodes that are relevant to neuropsychology. These include the *Neurology Exam Prep Podcast*, *Neurogenesis*, and the *Neurology Podcast*.

Ethics and Professional Practice

- Read the book, *Ethical Decision Making in Clinical Neuropsychology* (2018), by Dr. Shane Bush.
- Read the *Ethical Principles of Psychologists and Code of Conduct* (2017; https:// www.apa.org/ethics/code/).
- Review the Health Insurance Portability and Accountability Act (HIPAA) guidelines and details about your state laws (e.g., whether you are obligated to report an impaired driver).

- Review ethical information/questions from your EPPP study materials.
- Respond to mock ethics vignettes on BRAIN (https://brainaacn.org/mock-ethics-vignettes/).
- Discuss common ethical issues/dilemmas with your study group.

Pediatric Subspecialty Certification

ABCN offers a pediatric subspecialty certification to clinicians who have obtained board certification through ABPP/ABCN. According to the ABCN website, "the subspecialization process involves: (a) credential review of training, education, and practice in pediatric clinical neuropsychology; (b) written examination; and (c) submission of one subspecialty practice sample case" (https://theabcn.org/subspecialty-certification-in-pediatric-clinical-neuropsychology/). ABCN may offer additional subspecialty certifications in the future (e.g., in forensic or geriatric neuropsychology), but as of writing, the pediatric subspecialty certification is the only option.

Finally, there is another board that offers a certification in pediatric neuropsychology, namely, the American Board of Pediatric Neuropsychology (ABPdN). This board is not affiliated with ABPP or ABCN. For more information, visit https://theaapn.org/abpdn/.

Action steps
Ensure that you will be eligible for board certification by conforming to HC guidelines and ABCN regulations.
Once you begin independent practice, set aside clinical reports for cases that may be appropriate as practice samples.
To prepare for the written exam, read Stucky et al. (2020), as well as relevant peer-reviewed manuscripts and other neuropsychology books.
To prepare for the written and oral exams, join a study group and utilize the BRAIN resource.

Re-specializing in Neuropsychology

Up to this point, we have been covering traditional routes into neuropsychology for people in North America. To review, this is the process of earning a PhD or PsyD from an accredited clinical psychology program, completing a 1-year accredited psychology internship, and then finishing a 2-year neuropsychology fellowship that adheres to the HC guidelines. But not everyone traverses this path into our field, and we welcome people from a variety of different training backgrounds. This diversity will only strengthen our field, as it creates a wider set of perspectives, thinking styles, and problem-solving approaches. So if you are a clinical psychologist who was originally interested in specializing in trauma interventions, if you are a cognitive neuroscientist who is seeking to add patient care to your repertoire, or if you trained in an entirely different field (e.g., computer science, economics, marketing), it is still possible for you to become a neuropsychologist. The path can be challenging, and it is important to consider your individual circumstances when coming to a decision regarding whether or not to re-specialize.[29]

We will approach the topic of re-specialization (sometimes called "re-treading") by setting the bar at eligibility for ABPP through ABCN. You can review the eligibility criteria for ABPP in the Checklist below and ABCN's criteria in Ch. 6, but we will re-iterate the most important points here and address how to meet the criteria through additional training. You should know up front that re-treading is not as easy as merely completing extra CE courses or reading a few books. According to the HC guidelines, "Continuing education is not a method for acquiring core knowledge or skills to practice clinical neuropsychology or identify oneself as a clinical neuropsychologist. Continuing education also should not be the primary vehicle for career changes from another specialty area in psychology to clinical neuropsychology" (Hannay et al. 1998; p. 4–5).

Checklist for the ABPP general review:[30]

---

1) ___Doctoral degree (PhD, PsyD, EdD)[31]
2) ___The degree was from a psychology program which, at the time the degree was granted, was accredited by the APA, CPA, or an accrediting agency recognized by the US Department of Education.[32]
3) ___Your doctoral program met the requirements listed in the ABPP Generic Doctoral Program Eligibility document (www.NavNeuro.com/ABPPrequirements).
4)___You are licensed as a psychologist for independent practice at the doctoral level in a jurisdiction in the United States, its territories, or Canada.[33]

---

[29] If you decide that neuropsychology is not the right fit, there are several related fields that allow work at the master's level. Consider looking into training as a psychometrist, marriage and family therapist, social worker, school counselor, or research coordinator.

[30] This information is from the ABPP website (https://abpp.org/Applicant-Information/Degree-Requirements.aspx). If you are not sure whether you meet these requirements, reach out to ABPP (contact info is at the bottom of their website at https://abpp.org).

[31] According to ABPP: "Applicants who hold the Certificate of Professional Qualification in Psychology (CPQ) from the ASPPB qualify as meeting the doctoral degree and professional program requirements." For information, visit https://www.asppb.net/page/CPQ.

[32] If you obtained your doctoral degree prior to 2018, an acceptable degree is also one that is listed in the publication *Doctoral Psychology Programs Meeting Designation Criteria*.

[33] If you completed your doctoral degree and/or work outside of the United States, its territories, or Canada, then please contact ABPP to determine your eligibility status.

Every person considering re-specialization brings with them their own unique, nuanced story, and it is impossible for us to provide tailored advice to everyone. In lieu of that, we have three pieces of advice. First, we recommend that you write up a synopsis of your training/background and proposal for re-specialization (e.g., plan for obtaining didactics and fellowship experience) and submit this information to ABCN for advice and guidance prior to embarking on the journey.[34] Given the level of commitment required, you will want to know with a high degree of certainty that your re-specialization training will be sufficient to pass a credential review.

Second, we encourage you to document every experience related to neuropsychology and file away all written communications, syllabi, and other materials that may be helpful in order to demonstrate compliance with the HC guidelines. Third, we will sketch out a few general vignettes in order to illustrate potential paths to board eligibility. We think that one of these scenarios will resemble your own and will help provide you with a general idea as to your path forward. The following information pertains to individuals who received their doctoral degree after 2004.[35]

Scenario 1: You are a licensed psychologist who completed an accredited counseling or clinical doctoral program but took a generalist path, without any specialty coursework or training in neuropsychology.

This is the most common nontraditional scenario. Although the path is not necessarily easy, it is relatively straightforward. Given that you earned a clinically oriented doctorate and that you completed an accredited internship, you would not need to repeat these experiences. You are also a licensed psychologist, so you have already met the licensure requirements. Finally, you may also have some experience and training in psychological, psychoeducational, and cognitive testing. Presumably, the only area that is lacking from your training is in-depth, specialty training in neuropsychological populations and assessment methods. You would easily clear the ABPP general review but would not pass the ABCN specialty review. To receive clearance from ABCN, you would be required to complete a formal 2-year postdoctoral fellowship in clinical neuropsychology. The safest way to ensure that you meet ABCN's criteria would be to earn a position in an established fellowship program with a track record of producing board-certified neuropsychologists. However, any fellowship that met the HC guidelines would suffice. We suggest that you refer back to our section on the fellowship application process in Ch. 6.

---

[34] We recommend that you email the ABCN Credential Review Committee Chair (Credentialreview@theabcn.org).

[35] As noted above, ABCN training/eligibility requirements are directly tied to the year that the doctoral degree was conferred or the re-specialization program was completed. If you received your degree or re-specialized prior to 2005, see the ABPP website for the requirements that apply to you: https://abpp.org/Applicant-Information/Specialty-Boards/Clinical-Neuropsychology/Application,-Specialty-Specific-Fees.aspx.

It is difficult to say exactly what you could do to make yourself competitive to neuropsychology-specific fellowship programs at this stage. Because each program will differ, we recommend that you reach out to several training directors and supervisors at sites that interest you. It would likely impress them if you had already completed several neuropsychology-related courses such as those offered by the National Academy of Neuropsychology[36] or those offered at your local university. We also recommend that you read the seminal books (e.g., Lezak et al. 2012; Kolb and Whishaw 2015), regularly consume the neuropsychology literature, and attend multiple neuropsychology conferences (see Ch. 5).

Some people may wonder why a formal fellowship is necessary in order to call oneself a neuropsychologist. They might further contend that CE workshops and occasional consultation are sufficient to perform neuropsychological evaluations and take on the title, "neuropsychologist." As noted above, the HC guidelines explicitly prohibit this, and there is good reason for that. It would be tantamount to a primary care physician (PCP) calling themself a neurologist without completing a neurology fellowship. Yes, non-neuropsychologists may administer and interpret neuropsychological measures, just as a PCP may perform aspects of a neurological exam. However, there is simply too much nuanced specialty knowledge and too many skills required for someone to absorb all of it from occasional didactics or consultation.

Note: There are a handful of university-based programs specifically designed for re-specialization in neuropsychology. We cannot vouch for any specific program, but you can look into this as another option. APA lists a few such programs on their website: https://www.apa.org/ed/graduate/respecialization. However, keep in mind that a formal neuropsychological fellowship with on-site supervised experience is still required.

Scenario 2: You completed a nonaccredited counseling or clinical doctoral program (with a nonaccredited internship), and/or your doctoral program did not meet the requirements listed in the ABPP Generic Doctoral Program Eligibility document (see Checklist).

ABPP clearly states that your doctoral program must be accredited through APA, CPA, or an accrediting agency recognized by the US Department of Education and that it must meet the requirements listed in the ABPP Generic Doctoral Program Eligibility document (see Checklist). If your graduate program did not meet one of these criteria, we recommend reaching out to ABPP directly to ask about your options. The guidance laid out in the following document may also be of some assistance: https://www.apa.org/about/policy/chapter-9#respecialization-training. It is possible that you will be allowed to fulfill the requirements through

---

[36] Visit https://www.nanonline.org and then click "Continuing Education."

re-specialization at the graduate level (https://www.apa.org/ed/graduate/respecialization). However, this is not guaranteed. For example, some states do not approve people with re-specialization certificates for licensure, and you must hold a psychology license in order to pass the ABPP review.[37] If you cannot re-specialize, then, unfortunately, you will likely be required to complete an additional doctorate from an accredited program.

Although it is the case that doctoral programs must be accredited, this does not necessarily apply to internships. We could not find any materials specifically addressing this question on the ABPP general review pages. However, according to the ABPP specialty page for neuropsychology, "ABCN does not require neuropsychological training during internship. In addition, although ABCN prefers that applicants have completed an APA or CPA accredited internship program, the review decision regarding internship training is made by ABPP Central Office during the generic review. If ABPP passes the candidate's non-APA/CPA internship credentials during generic review, ABCN will typically defer to that decision" (https://theabcn.org/becoming-certified/). That being said, some states or provinces require an accredited internship in order to obtain licensure, so check your jurisdiction's requirements. Furthermore, many neuropsychological fellowships require completion of an APA or CPA accredited internship.

If you are able to clear the ABPP general review stage, it may still be difficult to attain approval through the ABCN specialty review process, given that many neuropsychology-specific fellowships require applicants to have completed substantial neuropsychological training at accredited doctoral and internship programs. To this end, we encourage you to follow the advice laid out in Scenario 1.

Scenario 3: You completed a doctorate in a nonclinical field of psychology (e.g., experimental), or in a field other than psychology, without an internship or other clinical experience.

In this situation, it is likely that ABPP would require you to complete a new clinical doctorate. The potential exception to this is for someone whose degree is in psychology; in this case, a re-specialization at the graduate level may suffice. If so, the guidance offered in Scenario 2 will apply to your situation.

****

No matter your situation, we understand that it is difficult mentally, emotionally, and financially to pivot to another career path. You may be required to uproot yourself and your family and move across the country. You will certainly be serving as a trainee and supervisee for longer than you otherwise would have. And you will be

---

[37] In addition to checking with your state/province's requirements, general licensure information can be found at https://www.asppb.net/page/guidelines; many jurisdictions use these guidelines as the basis for their standards.

waiting, possibly for years, before setting out as a full-fledged, independent practitioner. The sacrifices are substantial. But if neuropsychology is the career that will afford you a productive, fulfilling work life, then the cost is almost certainly worth it.

Action steps
Look closely at the requirements for both ABPP and ABCN approval, determining your unique situation.
Contact ABPP and/or ABCN if you are not sure whether you meet criteria or what the pathway to approval would be.

Conclusion

The roots of neuropsychology took hold in antiquity when humankind first began demonstrating an overt curiosity around the cause and structure of the mind and behavior. However, it wasn't until several millennia had passed and the scientific revolution had taken place that the tools became available to truly capitalize on this curiosity. Nineteenth- and early twentieth-century discoveries by neuropathologists and neuroscientists provided a biological basis for understanding the brain and paved the way for psychologically minded scientists to turn their attention to unlocking the secrets of the mind. Indeed, it was amidst the chaos and destruction of two world wars that neuropsychology began growing up and taking on its own identity. This occurred as a result of a confluence of movements, including intellectual assessment during WWI and WWII, advancements in statistical methods and psychometrics, and the release of early neuropsychological test batteries such as the Wechsler scales and the Halstead-Reitan battery. With this foundation in place, the first neuropsychology journal (*Neuropsychologia*) was established in 1963, the first professional society (INS) was founded in 1967, and the first two textbooks (*Neuropsychological Assessment* and *Fundamentals of Human Neuropsychology*) were released in 1976 and 1980, respectively. Since then, neuropsychology has undergone a revolution due to advancements in structural neuroimaging techniques, and it is in the process of undergoing another transformation today in response to managed care, genomics, and AI/informatics (Bilder 2011).

As a reader of this book, you are in the position to affect the future of neuropsychology. The field, while still small, continues to grow at a rapid pace and will likely be larger, stronger, and more innovative 10, 20, and 30 years down the road. By incorporating functional neuroimaging, computerized testing, data science, and biometric techniques, we hope to revolutionize the way in which the brain/behavior assessment and intervention are conducted. We hope that people who suffer from ailments such as dyslexia, Parkinson's disease, and traumatic brain injury will continue to turn to neuropsychologists for precise measurements, explanations, and diagnoses of their symptoms, as well as comprehensive treatment plans tailored to their individual characteristics and circumstances. But we will need hard work from a large number of bright, motivated teachers, clinicians, and researchers in order to further refine these services. We sincerely hope that you will consider contributing to what we feel is the greatest field ever: neuropsychology.

References

American Board of Clinical Neuropsychology (ABCN). (2020). *Board certification guidelines and procedures: Candidate's manual*. https://theabcn.org/becoming-certified/

Armstrong, K. E., Beebe, D. W., Hilsabeck, R. C., & Kirkwood, M. W. (2019). *Board certification in clinical neuropsychology: A guide to becoming ABPP/ABCN certified without sacrificing your sanity* (2nd ed.). Oxford University Press.

Bilder, R. M. (2011). Neuropsychology 3.0: Evidence-based science and practice. *Journal of the International Neuropsychological Society: JINS, 17*(1), 7–13.

Blumenfeld, H. (2010). *Neuroanatomy through clinical cases* (2nd ed.). Sinauer Associates, Inc Publishers.

Board of directors. (2007). American Academy of Clinical Neuropsychology (AACN) practice guidelines for neuropsychological assessment and consultation. *The Clinical Neuropsychologist, 21*(2), 209–231.

Bush, S. (2018). *Ethical decision making in clinical neuropsychology* (2nd ed.). Oxford University Press.

Cepeda, N. J., Pashler, H., Vul, E., Wixted, J. T., & Rohrer, D. (2006). Distributed practice in verbal recall tasks: A review and quantitative synthesis. *Psychological Bulletin, 132*(3), 354–380.

Cox, D. R. (2010). Board certification in professional psychology: Promoting competency and consumer protection. *The Clinical Neuropsychologist, 24*(3), 493–505.

Doran, J. M., Kraha, A., Marks, L. R., Ameen, E. J., & El-Ghoroury, N. H. (2016). Graduate debt in psychology: A quantitative analysis. *Training and Education in Professional Psychology, 10*(1), 3–13.

Dunlosky, J., & Rawson, K. A. (2015). Practice tests, spaced practice, and successive relearning: Tips for classroom use and for guiding students' learning. *Scholarship of Teaching and Learning in Psychology, 1*(1), 72–78.

Hannay, H. J., Bieliauskas, L. A., Crosson, B. A., Hammeke, T. A., Hamsher, K. deS., & Koffler, S. P. (1998). Proceedings of the Houston conference on specialty education and training in clinical neuropsychology. *Archives of Clinical Neuropsychology, 13*(2), 157–158.

Hansson, E., Mattisson, K., Björk, J., Östergren, P. O., & Jakobsson, K. (2011). Relationship between commuting and health outcomes in a cross-sectional population survey in southern Sweden. *BMC Public Health, 11*(1), 1–14.

Hessen, E., Hokkanen, L., Ponsford, J., van Zandvoort, M., Watts, A., Evans, J., & Haaland, K. Y. (2018). Core competencies in clinical neuropsychology training across the world. *The Clinical Neuropsychologist, 32*(4), 642–656.

Kelsky. (2015). *The professor is in: The essential guide to turning your Ph.D. into a job*. New York: Three Rivers Press.

Kolb, B., & Whishaw, I. Q. (2015). *Fundamentals of human neuropsychology* (7th ed.). Worth Publishers.

Künn-Nelen, A. (2016). Does commuting affect health? *Health Economics, 25*(8), 984–1004.

Lezak, M. D., Howieson, D. B., Bigler, E. D., & Tranel, D. (2012). *Neuropsychological assessment* (5th ed.). Oxford University Press.

Malhotra, D. (2014) also wrote a helpful article entitled, *15 Rules for Negotiating a Job Offer* in Harvard Business Review.

Malhotra, D., & Bazerman, M. H. (2007). *Negotiation genius: How to overcome obstacles and achieve brilliant results at the bargaining table and beyond.* Bantam Books.

Mamula, C., Barrett, B., & Mendosa, J. (2019). *ChooseFI: Your blueprint to financial independence.* Choose FI Media, Inc.

Ramsey, D. (2009). *The total money makeover: A proven plan for financial fitness.* Thomas Nelson, Inc.

Smith, G., & CNS. (2019). Education and training in clinical neuropsychology: Recent developments and documents from the clinical neuropsychology synarchy. *Archives of Clinical Neuropsychology, 34*(3), 418–431.

Stucky, K. J., Kirkwood, M. W., & Donders, J. (Eds.). (2020). *Clinical neuropsychology study guide and board review* (2nd ed.). Oxford University Press.

Sweet, J. J., Klipfel, K. M., Nelson, N. W., & Moberg, P. J. (2020a). Professional practices, beliefs, and incomes of US neuropsychologists: The AACN, NAN, SCN 2020 practice and "salary survey". *The Clinical Neuropsychologist,* 1–74.

Sweet, J. J., Klipfel, K. M., Nelson, N. W., & Moberg, P. J. (2020b). Professional practices, beliefs, and incomes of postdoctoral trainees: The AACN, NAN, SCN 2020 practice and "salary survey". *Archives of Clinical Neuropsychology.*

Tyson, E. (2019). Personal finance for dummies (9th ed.). John Wiley & Sons, Inc.

Whiteside, D. M., Guidotti Breting, L. M., Butts, A. M., Hahn-Ketter, A. E., Osborn, K., Towns, S. J., et al. (2016). 2015 American Academy of Clinical Neuropsychology (AACN) student affairs committee survey of neuropsychology trainees. *The Clinical Neuropsychologist, 30*(5), 664–694.

Yerkes, R. M., & Dodson, J. D. (1908). The relation of strength of stimulus to rapidity of habit-formation. *Journal of Comparative Neurology and Psychology, 18*(5), 459–482.

Index

© Springer Nature Switzerland AG 2021
J. A. Bellone, R. Van Patten, *Becoming a Neuropsychologist*,
https://doi.org/10.1007/978-3-030-63174-1